GRAPHENE

GRAPHENE

THE SUPERSTRONG, SUPERTHIN, AND SUPERVERSATILE MATERIAL THAT WILL REVOLUTIONIZE THE WORLD

LES JOHNSON AND JOSEPH MEANY

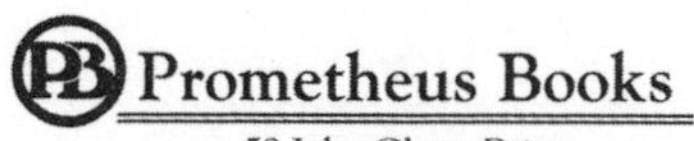

59 John Glenn Drive
Amherst, New York 14228

Published 2018 by Prometheus Books

Cover design by Jacqueline Nasso Cooke
Image of graphene sheet © Graphene Supermarket
Image of glass © laboratory / Alamy Stock Photo
Image of activated carbon powder mound © PictureLake / Getty Images
Image of graphene on flower © Long Wei / EPA / Shutterstock
Image of 3-D graphene model © Melanie Gonick / MIT
Image of solar panels courtesy of Larry E. Reid Jr. / US Air Force
Image of eye © stefano carniccio / Alamy Stock Photo
Cover design © Prometheus Books

Inquiries should be addressed to
Prometheus Books
59 John Glenn Drive
Amherst, New York 14228
VOICE: 716–691–0133 • FAX: 716–691–0137
WWW.PROMETHEUSBOOKS.COM

22 21 20 19 18 5 4 3 2 1

Library of Congress Cataloging-in-Publication Data Pending

Printed in the United States of America

Les Johnson:

This work would not have happened if it weren't for the support, love, and boundless enthusiasm of my wife, Carol, who also introduced me to my coauthor, Joe, in the halls of DragonCon. Thank you, Carol, for being there for me in our first thirty years of our marriage. You are loved.

Joseph Meany:

Firstly, I would like to dedicate this to my parents, Mark and Sharon Meany. I have them to thank for my curiosity and passion for the wonders of the natural world. I'd like to also dedicate this to my many mentors and teachers throughout my scientific growth, particularly Mr. Robert Shalit and Mr. Joseph Puleo for seeding my love of chemistry while at Keene High School.

CONTENTS

PART FOUR: WHAT'S NEXT?

PREFACE

What if you discovered an infinitesimally thin material capable of conducting electricity, able to suspend millions of times its own weight, and yet porous enough to filter the murkiest water? And what if this substance was created from the same element as that filling the common pencil? This extraordinary material, graphene, is not a work of science fiction. A growing cadre of scientists aims to make graphene a mainstay technological material by the second half of the twenty-first century. Not satisfied with that timeline, some entrepreneurial types would like to see widespread adoption of graphene within the next decade. How could this be possible?

Graphene is elegant. It is created from a single element, carbon, formed by just one type of bond. Despite graphene's apparent simplicity, isolating the material was an elusive "Holy Grail" for chemists and physicists alike. Even as the periodic table extended beyond the hundred-odd elements naturally found on Earth, galaxies were charted, and the human genome solved, this material, with the simple chemical formula of C, remained a distant goal at the frontiers of science. Why was this? Graphene excels at hiding in plain sight, and the techniques and instrumentation perfected in the last two decades have played a pivotal role in its discovery.

Carbon, the sole constituent of graphene, is all around us. The element is the fourth most common in the entire universe. Most people think of materials in terms of atoms and molecules, where molecules are made from defined types and numbers of atoms. With graphene, counting carbon atoms is inconsequential. Merely the way in which the constituent carbons are bound to one another is crucial, with this feature separating graphene from other wholly carbon materials like diamonds and graphite. At the atomic level, the exclusively carbon graphene resembles a hexagonal "chicken wire" fence, with each carbon atom making up the point of a hexagon. The hexagonal distribution makes graphene's earth-shattering properties possible, as the distribution allows the individual carbon atoms of graphene to lay flat.

This property of graphene cannot be overlooked. Graphene is a perfect anomaly in the world of chemistry—a flat, two-dimensional molecule, with a single sheet of graphene measuring only one atom thick. You might immediately question the structural integrity of graphene due to its delightfully simplistic construction, but the weaving of the carbon hexagons throughout the structure makes the atomically thin material unexpectedly strong.

Proper application of graphene holds the key to revolutionizing materials technology in the latter half of the twenty-first century, but at what cost? Thankfully, not a substantial environmental one. There is a critical difference between graphene and another linchpin of modern technology, rare-earth metals. These hard-won rare-earth metals, metals including tantalum, neodymium, and lanthanum, are found everywhere, from the inside of our smartphones to pharmaceuticals. Unlike with rare-earth metals, we do not need armies of manual laborers assisted by heavy equipment and an endless parade of fifty-five gallon drums of polluting solvents to find and retrieve graphene, due to one simple fact: graphene's elemental constituent, carbon, is all around us. The most common precursor of graphene today is the mined mineral graphite. Rare-earth metals are scarce, but the integration of graphene into our lives would not be driven by the acquisition of raw materials and disputes between superpowers, but would be guided by the possession of knowledge, with patents and technology separating the victors and the vanquished.

You have experienced synthesizing graphene, maybe even earlier today, on a very small scale. The pressure exerted by your hand and fingertips likely created a few layers of graphene the last time you ran a pencil across a notepad, turning humble graphite into graphene as you wrote this week's grocery list. But if graphene can be made by such simple means, and its sole constituent, carbon, leads oxygen, nitrogen, and hydrogen in the hierarchy of elements that construct our living world, why is graphene just now, in the twenty-first century, coming to the forefront of human understanding?

The answer to this question is where the story resides. The story of graphene is a story of accidental discovery. A story of corporations and governments racing to spend billions of dollars in hopes of funding research and development projects to discover a material still years away from store shelves. A story of new materials that will disrupt the way we create things, and, in doing so, what we can create. The previous technological revolutions taught us many things. Each new discovery allowed us to break into new experimental territories and further our understanding of what is possible to accomplish. Chemical batteries allowed energy to be stored for future use (like light at night). Steam power allowed us to generate tremendous amounts of energy to accomplish tasks no living thing could. This new revolution may allow us to throw off the shackles of metallic wires.

If you are curious about science, economics, history, or the vague point where all three of these topics overlap, then you will probably enjoy this book. If you already know what graphene is, then you might wonder where and why history might play into such a recent discovery. After all, graphene as a material for the future has only been in the news for about ten years.

Since at least the 1950s, people have been trying to take graphite out of the ground and turn it into a pile of black gold. This effort was met with fifty years of resistance from the graphite, which has not so easily been coaxed to divulge its secrets. When graphene was finally isolated and examined, physicists and chemists were astounded at what they found. The history beneath this discovery is not so straightforward, though, and it traces its roots all the way back to 1859 in Great Britain. How appropriate, then, that the country already well-known for its history involving carbon should be the country where single-layer graphite was finally witnessed.

After two researchers in Great Britain, Konstantin Novoselov and Andre Geim, were awarded the Nobel Prize in Physics in 2010, technology magazines everywhere heralded a new era of "wonder materials" based around this atomically thin tessellation of carbon atoms. With its incredibly high strength and almost impossibly low electrical resistance, graphene pulled back a hidden curtain, allowing scientists to catch a glimpse of the marvels that lay beyond. With the shrouds lifted, the groundwork was laid to revolutionize how we will go about designing and making everything from cars to vaccines and from food packaging to spaceships.

The economic potential of this material cannot be understated. Being atomically thin, graphene can be incorporated almost seamlessly into any modern product, with appreciable effect. Early investors were burned, however, by entrepreneurs who over-promised and under-delivered on performance aspects for products (especially composites like plastics) that had graphene in them but that did not use graphene in a way that made its incorporation worth the added expense. It was, in some cases, just an added bit of snake oil. As the overall volume from new production methods and the quality of the resulting graphene have both increased with time, we are starting to finally see graphene's true benefits. Governmental support is higher than ever in many countries, as whomever discovers a high-throughput production method for pristine graphene will reap significant financial rewards on the world stage.

Part One

DISCOVERY AND CONTROVERSY

Every new discovery may be considered as a new species of manufacture, awakening moral industry and sagacity, and employing, as it were, new capital of mind.

—Humphry Davy, *Edinburgh Review, or Critical Journal: For June . . . October 1827*

Chapter 1

CARBON, CARBON, EVERYWHERE!

Perhaps the second oldest trope in chemistry is, "Don't trust atoms, they make up everything." It's funny in that double entendre way that atoms do compose every bit of matter in the known universe, and that they are lying little buggers.

This may come off as laughably obvious, but you're holding an object in your hands at the moment.[1] Whether you are reading this as a physical book, on an e-reader, or on some other digital device, there is *something* in your hands. The construction may vary; books don't seem to have much in common with digital devices. Regardless of the materials in the object itself, though, the important point is that it is made of matter. But what does it really mean that there is matter? Why does that, frankly, matter?

The materials that make up whatever is in your hand are formed from atoms. Atoms have many different types of names, and I'm not talking about Phil, Anne, or Charley. One type of atom, with a specific set of properties, might be called argon. Another might be called tungsten. A third might be called carbon. What is in a name? We'll get to that in a minute. Elements, which are atoms all of the same type, are the tools that chemists work with to create glue, plastic bottles, medicine, food, and everything you can imagine. You're probably familiar with oxygen. We need it to breathe. It's in water, glass, rocks, and many drugs. You are probably familiar with iron, too. It's in cookware, tools, and even your blood. Helium, iron, and oxygen, are all examples of elements.

Episode nine from Carl Sagan's 1980 series *Cosmos*, "The Lives of the Stars," opens with an apple on the screen suspended against a black outer space–like backdrop. Suddenly, a knife slices the apple in two, and the scene moves to a baroque dining hall where Sagan (the host) is being served an apple pie.

The apple reference from *Cosmos* is a tip of the hat to a Greek philosopher Democritus (sometimes spelled *Demokritos*) of Abdera, who, along with his mentor Leucippus (or *Leukippos*), developed the earliest atomic theory around 450 BCE. As the story is told, they developed the idea by imagining a knife cutting an apple in half. You can cut that half into two halves again, giving two quarters. But they went further. How many times can you halve an apple? By imagining an impossibly sharp knife, they wondered whether continued cuts on the apple would eventually cause the apple to lose its identity. In other words, where does identity begin and end within a material, or is there a transition at all? This concept was a particularly huge callout against two other philosophers of the time discussing atomic theory, Aristotle and Anaxagoras. Aristotle and Anaxagoras argued that no matter how many halves one cut, an apple would always be an apple and a gold nugget would always be a gold nugget. No matter how small in the universe you zoomed in with your magnifying glass, you would always be able to tell apart two substances from one another. This assumption imbued a sort of inherent quality to every single thing in existence. It instilled a permanence and order to the universe, which Aristotle attributed to divinity, a quality that quite obviously appealed to religious opponents of atomic theory for centuries to come.

Democritus and Leucippus didn't like the blatant crutch of divinity in Aristotle's argument. They suggested that objects are made up of some strong, uncuttable material that exists within some sort of empty space or *void.* The idea of a void was unusual at this time, as humanity had no concept of what lay beyond the atmosphere. All that philosophers knew led them to believe that the sky extended all the way to some crystal sphere. "Outer space" and a vacuum were outside the common wisdom. But, for Democritus and Leucippus to be right, there needed to be some sort of space for the particles to move around. For movement to occur, particles had to displace and replace one another as in a fluid. In an attempt to extend their analogy, a ship "cuts" through the water as a knife through the apple. In order to make headway, the prow must push water out of the way while water fills back in with the wake, and the knife pushes apple out of the way while air fills in the gap. Eventually, though, this impossibly sharp knife would have to hit something that it couldn't cut. This indivisible part, this thing that could not be cut, Democritus called an *atom*. The word derives from the Greek "*a–*" meaning "not," and "*–tomos*" meaning "to cut." These atoms could form the building blocks on which many different materials could be made without a creator having to devote individual attention to all things within the universe. We now have these particles that can't be divided, called atoms. Democritus and Leucippus came up with a less catchy stance about duplicitous particles "making up everything." To Democritus and Leucippus, only two things existed—atoms and the empty and nearly endless void that they populate. It is a fundamental principle of the atomic theory that atoms are indestructible particles. It wasn't until the twentieth century that Henri Becquerel, Marie Curie, and Pierre Curie discovered that even atoms may break, though through a process far beyond the imagination of early natural philosophers and researchers. However, atoms as elements are still fundamental in one way—once the atom has been broken into its constituent parts, the elemental identity of the atom is lost. Therefore, from a certain perspective, atoms are still uncuttable.

Concurrently with the developments in Ancient Greece, Indian philosophers were also writing works that related to speculation about the fundamental nature of the universe. Pakhuda Kaccayana and Kanada were two early Indian proponents of atomism in Eastern culture.[2] They, too, faced some criticism from their contemporary colleagues. Constancy within the material realm was proof (to the opponents of atomic theory) that creation was a product of divine inspiration, and that a breakdown of this principle would mean a breakdown of divinity itself, along with most fundamental religious positions—most importantly the loss of eternal salvation. Most arguments against ancient atomism stated that if atoms are eternal and irreducible, then they do not allow for some sort of soul that passes to a holy realm. This, clearly, did not play well with early Christianity (which also had a tremendous issue with the mathematical concept of 'infinitesimals') as well as with other theist practitioners. It wasn't until the Islamic Golden Age (~700 CE–1200 CE) that new developments in atomic ideas began to seriously take root. Avicenna[3] and Averroes[4] were two Muslim scholars who were able to merge Indian and Greek philosophy into coherent ideas that took root throughout Europe and Southeast Asia.[5] As a testament to the quality of their contributions, Avicenna's writings greatly influenced two early physicians—Franciscan friar Roger Bacon (*Doctor Mirabilis)* and Saint Albertus Magnus.[6]

Despite the growing popularity of the idea of fundamental indestructible particles, actual experimental proof of atomic particles and their behavior eluded investigators until Robert Boyle published *The Skeptical Chymist* in 1661. Within that same book, Boyle dismissed the Aristotelian "elements" of antiquity—fire, water, air, earth, and ether—in favor of chemical elements more like those we would recognize today. Isaac Newton, best known for his pioneering work in mathematics and physics, concurred with Boyle's findings.[7] These two great minds differed on a significant point, however. Boyle mostly dismissed the alchemical arts while Newton embraced them. Boyle and Newton, along with Descartes (of "I think, therefore I am" fame), Pierre Gassendi (a French scientist-priest), and Roger Joseph Boscovich (a Ragusan scientist-priest), laid considerable groundwork for the discovery of all 118 modern chemical elements.[8] The 1700s and 1800s were a period of unprecedented discovery, where new elements were finally being discovered, disrupted from their natural minerals and ores at a pace not seen before or since. Periodic trends began to emerge when a Russian scientist, Dimitri Mendeleev, constructed the first primitive periodic table of the elements in 1871.

Finally, it was Ernest Rutherford who was able to deduce through a series of experiments during 1908–1910 that atoms weren't simply tiny solid balls of matter. Rutherford made an apparatus that fired alpha particles—which are basically helium nuclei stripped of their electrons—at a sheet of gold foil. Most of the particles passed through the foil with only a small bit of deflection from their original trajectories. The surprising result was that some particles bounced in completely different directions. A few particles even bounced back at the gun in an extraordinary rebound. At first, the result confounded Rutherford and his coworkers. This was the first time that anyone had experimentally witnessed that atoms are mostly empty space with a tiny but incredibly dense center. A vision of atoms was finally coming into focus. Think of the philosophical implications from Democritus's point of view. He had said that there were two things in the cosmos: atoms and the void. The kicker from Rutherford's findings is that atoms are mostly void, too.

As we have come to understand them in the twenty-five hundred years between Ancient Greece and modern times, atoms are made up of three basic parts. There are *protons* and *neutrons* that glom together in the nucleus and give the atom its weight. The number of protons, as mentioned above, gives an atom its identity. The number of neutrons, however, can vary and still not change the atomic element. Changing the element can only be done by changing the number of protons in the nucleus. An atom with seven protons and seven neutrons is nitrogen-14, referring to the sum of the masses of the protons and neutrons. If an atom has seven protons and eight neutrons, it is still nitrogen but is heavier than nitrogen-14. It is nitrogen-15. Beyond the nucleus, *electrons* are spread out in a diffuse cloud, zipping far out in patterns called shells or orbitals. This cloud gives an atom its volume, although they contribute an almost negligible amount of mass. While ninety-nine percent of an atom's mass comes from the nucleus, the nucleus is like a pea in a football stadium otherwise filled by the electron cloud.

Consider this: a teaspoon of butter weighs about six grams—something you can obviously hold very easily in your hand. However, if you had a teaspoon of nucleus matter (just the protons and neutrons) stripped of their electrons, then your spoon's contents would weigh as much as a very large mountain! You, me, this book—we're all just empty space with a few truly solid bits thrown in for good measure. It looks like Democritus had the right idea after all.

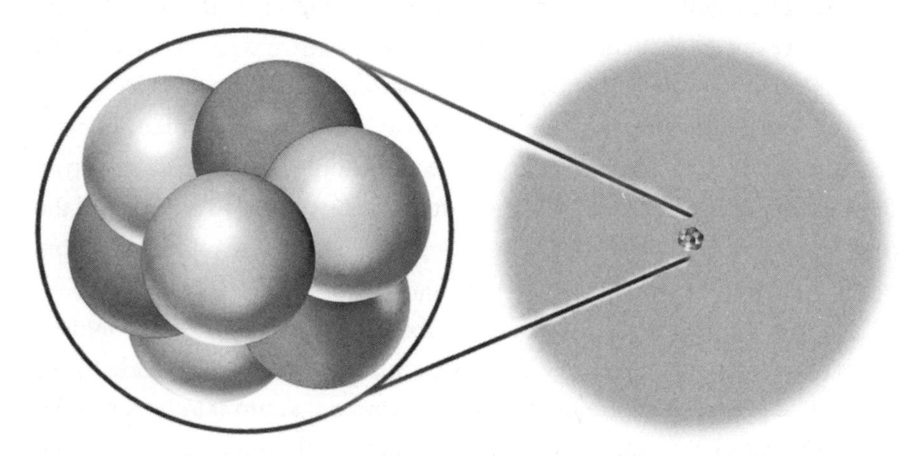

Figure 1-1: An artist's representation of an atom. Protons (dark gray) and neutrons (silver) are contained within the nucleus. The diffuse electron cloud dwarfs the nucleus itself. (Image by Joseph Meany.)

Books contain lots of different types of molecules: long chains of sugars linked together to form the solid starches in the pages, sticky adhesives in the glues, and different dyes that make up inks for marking the pages. Digital devices are much more complicated. They contain circuits made from metal and ceramics, screens made from glass and special dyes, and protective cases made from plastic and metal. Molecules lead to the seemingly endless possibilities behind all of chemistry which underlies medicine, technology, biology, and material science.

What are these things we call metals and sugars? How are they different and how are they basically the same? At first glance, the question and the answer seem absurd. Tools are made from metal, and sugar goes in your coffee, right? It can't possibly get more complicated than that—can it? Yes, it can. Metal atoms sit next to each other in ways that allow them to slide past one another when enough heat and pressure are applied.

This is how a blacksmith beats iron into a sword or a large machine turns aluminum blocks into the foil used to wrap your leftovers. Electrons interacting between metal atoms do so in a way that lets the electrons move about easily. They slosh around almost freely, which is why metals are such useful conductors of electricity. Electrons binding silicon and oxygen atoms in glass are connected much more intimately; glass is not as malleable as metal and tends to be brittle. You've seen this difference in properties if you've ever dropped a metal cup (which will bounce or perhaps dent with enough force) or a glass cup (which obviously shatters). The difference is due to the fact that electrons shared between the silicon and oxygen in glass are not freely distributed around the atoms. Rather, they are kept between the two atoms. This *localization* is why glass doesn't conduct electricity. *Delocalization* of electrons, on the other hand, is the name for the phenomenon that causes metals and other conductors, like graphene, to work as they do.

The particular arrangement of atoms in specific shapes is what gives rise to the properties of a molecule. Think of how specific shapes make up a house. There are lots of ways to build a house, although with a generally limited selection of materials with which to construct it. This is why houses can look so very different. The specific arrangement of the building materials, though, is what makes a house *your house*. It is the shape of the house that gives the house its identity as either yours or mine. Likewise, arranging carbon, hydrogen, oxygen, or other elements in different shapes will give you glucose, aspirin, or acetone.

Of course, this was not always so well understood. As the Middle Ages dragged on, royal philosophers broadened the scopes of their inquiries. No longer were court astronomers and mathematicians limited to charting the movements of stars or cataloging crop yields just for tax purposes. Among their duties, some early investigators involved themselves in the protoscience of alchemy. While the chemical principles behind smelting ores and making ceramics had been known since at least 3000 BCE, more complex and specific knowledge about these processes did not come about until people began experimenting in repeatable ways and reporting the results of the experiments to their colleagues.

The chief motivation behind these experiments was often alchemy, the desire to turn lead or mercury, so-called "base metals," into coinage metals like silver or gold. This process was called *transmutation*. As we mentioned earlier in the chapter, elements are defined by the protons in their nuclei, which are untouched during chemical reactions. Nevertheless, alchemy in the Middle Ages evolved as a practice with the assistance of the Catholic Church from about 1200 CE to the mid-1600s. Many skilled artisanal jobs, such as professional smiths, apothecaries, and other chemically oriented professions were developed during this period. One type of artist especially flourished in this environment—the con artist. Alchemy's obsession with finding the formula for transmutation gave unscrupulous individuals a new way to sell "miraculous cure-alls" and get-rich-quick schemes to unsuspecting townsfolk:

> Nothing is more astonishing than that persons should be found credulous enough to be the dupes of such impostors. The very circumstance of their claiming a reward was a sufficient proof that they were ignorant of the secret which they pretended to reveal; for what motive could a man have for asking a reward who was in possession of a method of creating gold at pleasure? To such a person money would be no object, as he could procure it in any quantity.[9]

Early alchemists developed get-rich-quick schemes by performing demonstrations for potential patrons. These demonstrations involved some sort of deception on the part of the "experimentalist," such as heating crucibles with false bottoms to reveal gold. Patrons would then pay a steep reward to the alchemist for sharing his methods. The patron would only realize the deception after returning home to attempt the transmutation themselves.[10] There, the unfortunate patron would find that they had been duped. Another method for supposedly transmuting base metals into gold came from using trick nails. These false nails had gold or silver soldered onto the iron, and the nail was then covered in some sort of ink or other obscuring substance that could be dissolved away when the nail was dipped into a special chemical solution. The hidden gold would be "revealed," to the wonder and amazement of the patron.[11]

Robert Boyle, though, effectively transformed alchemy from a profit-seeking scam art to an investigative science.[12] In *The Skeptical Chymist*, Boyle is careful to refer to elements as irreducible parts or pure substances.[13] His careful experiments and observations recorded that more complicated substances—rocks, plant matter, gasses, etc.—were able to undergo chemical reactions and separation from one another. More importantly, these reactions and separations were predictable and repeatable, untethered to the will of the gods or some other mysterious magic. While gold was the most common alchemical trick to perform during the early days of chemistry, the discovery of phosphorus in 1669 allowed for a new explosion of pseudoscientific demonstrations for a time. Transmutation of elements was not witnessed until the late nineteenth century, when Becquerel and the Curies discovered radioactivity from the decay of atomic nuclei. Fusion, another method of nuclear transmutation, was not developed until well into the twentieth century, with the development of thermonuclear weapons. Today, manmade nuclear fusion (outside of the natural environment within the hearts of stars, stable or exploding) is the source of all elements above atomic number ninety-five.

This concept—that there are pure substances that undergo reactions with one another to form more complicated structures—is the entire foundation of chemistry as a science. Elements make up molecules and molecules make a pair of scissors, a cheesecake, or a cat.

One large section of the puzzle was left to John Dalton to solve, and in 1803 Dalton determined from experiments that samples of these pure atomic substances were able to combine and form what he called "compound atoms."[14] Water, carbon dioxide, nitric oxide, sulfuric acid, and others are all examples of compounds that Dalton focused on. The key finding from this, and later work that refined his hypothesis, is that each compound must contain a specific proportion of elements to one another. Eventually, this came to be represented in the molecular formulas that many of us are familiar with today—H_2O, CO_2, NO, H_2SO_4. These formulas tell us about the chemical by the ratios of elements to one another. In water, H_2O, there are two hydrogen atoms for every one oxygen atom in the molecule. In carbon dioxide, there are two oxygen atoms for every one carbon atom, and so on.

For most of recorded history, people regarded the chemistry of living things and nonliving things differently. Rocks and minerals were clearly different from living things. Organisms contained fats, proteins, sugars, and oils which are all carbon-based molecules. Thus, the study of chemistry from living (carbon-containing) systems was labeled organic chemistry. Chemical systems that dealt with molecules that were not derived from something living logically fell under the umbrella of inorganic chemistry. People believed for a long time that the two branches of chemistry were entirely separate and that organic molecules contained some sort of vital life force that made them distinct from inorganic molecules. As such, chemicals of natural or biological origin were simply assumed to be completely incompatible with inorganic molecules. This was reinforced by the invisible connection between food cycles—living things consume only other living things for nutrition and supposedly leave the inorganic soil unchanged. It was unthinkable that something alive would choose to eat rocks, after all.

The assumption that living things required a special divine spark, called vitalism or the vitalist doctrine, was turned on its head in 1823 when a twenty-three-year-old medical doctor from Germany named Friedrich Wöhler evaporated a solution of ammonium cyanate (NH_4OCN) in water.[15] He expected to get the salt back out again, but was met with a curious surprise: the inorganic salt had transformed into a different molecule, urea. You know this as one of the primary components of urine.

In fact, Wöhler was also the first person to make a urine-related pun (in a letter to his postdoctoral advisor, no less). He wrote,

> In a manner of speaking, I can no longer hold my chemical water. I must tell you that I can make urea without the use of kidneys of any animal, be it man or dog.[16]

This excitement eventually coalesced into the knowledge that any molecule can be synthesized by humans if it has been made by nature. This also means that, despite the perceived origin of the atoms within that molecule, all atoms of a particular element have the exact same properties. Carbon released from carbonates trapped by primordial oysters is the same as carbon trapped in oil fields and is the same as carbon being excreted in the bathroom.

When we consider chemicals and their properties, bonding is king. The elements involved play an important role, of course, but it is important to remember that chemical reactions and bonds are the economy of electrons. For instance, what is the difference between coal, graphite, and diamond? If you look at samples on a table, you would probably be able to name plenty of differences right off the bat.

Coal is jet black, inconsequentially light, and brittle. Diamond is popularly familiar to everyone as well. It's clear and colorless when polished and incredibly hard. Graphite as a lump is a dull, lustrous gray material that looks almost metallic. Powder it, though, and you would be hard-pressed to tell the difference between coal dust and graphite dust using eyesight alone. Slightly related to graphite is the fullerene class of molecules. Fullerenes don't look like much; its particles are small, the powder is very fine and light, and it is soft to the touch.

These materials are so utterly different in their properties that, without prior knowledge of their composition, it would stretch the imagination to find something in common between them all. Underlying the cosmetic differences between these three materials, however, is the element carbon. Carbon atoms, all having six protons and between six to eight neutrons in the nucleus, connect to each other in different ways. This gives rise to the properties we see in brittle coal versus soft graphite versus lustrous diamond. We do not know exactly when carbon was discovered, but its importance as a fuel for fire in the form of sticks and other dead organic materials surely hints that it was recognized as a substance around the time that humans tamed fire for our own uses. After that, coal's reactivity was used for smelting metal ores dug out of the ground to produce shiny metal jewelry and weapons. These different forms of elements that differ only by the way the atoms are connected to one another are called *allotropes*. Diamond forms cubes of atoms, graphite/graphene forms sheets, and the fullerenes form balls. It is the allotropic form of an element that decides the properties we witness on a familiar scale. The cubes of carbon atoms are what make diamonds so rigid and hard. It is the plates of graphite that make it smooth, lubricating, and flexible (if you're only considering a single sheet).

When carbon as an element is compared to the periodic table as a whole, it could almost be considered boring. Unlike the bottom of the table, it's completely stable and nonradioactive. It's uninteresting to pyrotechnics enthusiasts, who much prefer the colors afforded by alkali and alkaline earth metals along the left-hand side of the table. You can't cast it to make weapons or machinery as you can with iron. It isn't particularly pretty (except in diamonds) which makes it not nearly as covetable as the coinage metals copper, silver, and gold.

Carbon sits toward the right-hand side of the periodic table, in what chemists call the *p*-block. It's a lightweight, unassuming element that doesn't draw the eye like liquid mercury or inspire visceral fear as with uranium or plutonium. Even purple iodine has a shock-and-awe advantage over that dirty lump of coal that no child wants to receive in their holiday stocking.

However, carbon is special in its mediocrity. The bonds that it does form are strong enough to hold together most molecules over the temperature range to which our planet is subjected. And yet, the bonds are not so strong that its chemical reactions are a one-way street. Aided by energy from the sun, plants are the wardens of the carbon cycle whereby life-sustaining chemistry is recycled and replenished. Proteins are recycled for many repetitive chemical reactions within cells, while the chemicals those proteins work on (smaller molecules) must be replenished by our food (plants, ultimately).

But simple carbon, the sixth element, is exactly what makes life and all that we understand to be "alive" possible. Through its ability to share its four outer electrons, to make a maximum of four bonds to other elements, carbon, in the form of graphene, is poised to bring about a new era and replace silicon as the dominant element in our technological society.

The ability of carbon to make a total of four bonds is more important than it seems at first glance. Why four, and not something like three or five or twelve? Why are four bonds even that important? To understand that, we need to focus on the electrons of an atom. Remember that, as atomic number 6, carbon has six positively charged protons in the nucleus. To balance this +6 charge, we need six negative charges from electrons. Thus, carbon has a total of six electrons.

"But wait," you ask, "didn't you just tell me that carbon makes four bonds, not six?" This is a fair question. Two of those six electrons are closer to the nucleus than the outer four, and are therefore unavailable to make outside bonds. This is because electrons arrange themselves in shells, or *orbitals*, that have distinct sizes and shapes. The smallest shell holds two electrons, and so the extra four are available to bond with up to four other electrons. This attribute was a hot button topic in the early 1850s. The world's best chemists, from London to Darmstadt, were steeped in fervent debate at this time, shooting letters across the continent to figure out how and why atoms come together to form molecules. They understood that molecules come together in specific proportions, but the shape of molecules and the ways atoms connect to one another remained elusive.

In 1854, August Kekulé was on the way home from having dinner with a friend when he dozed off during his carriage ride. He recalled later,

> On a fine summer's evening, I was travelling once again with the last omnibus through the then deserted streets of the metropolis, usually so full of life,—"outside" on the top deck of the bus, as usual. I fell into a

> reverie. Then the atoms gamboled before my eyes. I had always visualized them in motion, those little beings, but I had never succeeded in discovering the nature of their motion. *To-day, I saw how two small ones often joined up to form pairs, how larger ones seized two small ones, still larger ones kept hold of three or even four of the small ones, and how everything revolved in a whirling dance. . . .* The conductor calling "Clapham Road" wakened me out of my dream, but I spent a part of the night putting sketches at least of that dream picture on paper. *Thus arose the theory of structure.*[17] [emphasis added]

If you've ever taken a high school or college chemistry class, you may remember the words "octet rule." This is the idea that atoms will seek to fill up their outermost shell with eight electrons by sharing electrons with other atoms. The noble gases, like neon or argon, already have eight electrons in their outer shell, which is why they do not react with other atoms to form molecules. Halogens, like chlorine, tend to form a single bond, due to their seven electrons in the outer shell, and oxygen atoms form two bonds because of the six electrons in their outer shell. The shapes of molecules are determined by how many bonds a particular atom can make. These shapes are a critical determining factor in the ability of electrons to move about a structure, which, as we described earlier, is the very essence of conduction.

Now, since carbon's outer shell contains four electrons out of a maximum of eight, it can form bonds with up to four different atoms. These bonds need not be evenly distributed between four atoms, however. Single bonds between carbon and an atom to which it is connected concentrate electrons in the space immediately between the two atoms. The electrons are held static (in a sense), and are therefore called *localized electrons.* When carbon is able to form multiple bonds to a single atom, however, something special happens. The second bond between carbon and its neighbor means that the electrons in that double bond are no longer specifically located between the atoms. Rather, they are spread out in space; the orbital is far more diffuse. They are *delocalized.* Remember back a few pages when we talked about delocalized electrons moving around a sample, leading to electricity flow? If strings of carbon–carbon double bonds were connected in a row, electrons could move back and forth across the carbon atoms like they would in a wire. In fact, that is exactly the idea behind several fields of research these days. Scientists want to create molecules out of carbon using structures with lots of delocalization in them to create a series of wires and other computer components. This field is just beginning to catch on and has been dubbed "molecular electronics." We discuss molecular electronics in greater detail in the next chapter.

There is a molecule of carbon and hydrogen that involves these multiple bonds and that was especially important in figuring out how the delocalization of electrons affected organic molecules. This molecule might be something you're familiar with as a component of gasoline—it is called benzene.

As other chemists were hard at work trying to deduce the structures of molecules, Kekulé decided it was time to take a nap:

> I was sitting there working at my text-book [*sic*]; but I made no progress—my mind was on other things. I turned my chair to the fire and dozed off . . . the atoms gamboled before me. This time, small groups remained modestly in the background. My mind's eye, sharpened by repeated visions of a similar kind, was now able to distinguish larger structures of many varied arrangements . . . and look—what was that? One of the snakes gripped hold of its own tail and mockingly the structure whirled around before my eyes. As if by a flash of lightning I awoke.[18]

The snakes, as the sleepy scientist realized, stood for the six-sided ring formed by the carbon atoms contained within the molecule. The vision of a snake biting its own tail is hardly an accident. One of the most enduring symbols from the alchemical era was the *Ouroboros*, a snake in a ring devouring its own tail, a symbol of eternal creation and destruction. Eventually, Kekulé published a structure for benzene that considered the need for each carbon to have four total bonds (two to an adjacent carbon, one to another adjacent carbon, and one to an attached hydrogen). In this structure, the double bonds are staggered with single bonds in a 1-2-1-2-1-2 arrangement. Kekulé did not live to see experimental proof of his prediction, as he passed away in 1896. In 1928, he was finally vindicated when E. Gordon Cox confirmed the crystalline structure of benzene. Cox demonstrated that all the carbon–carbon bond lengths are the same in the structure—perfect symmetry. A few years later, a London scientist named Kathleen Lonsdale looked at the crystal structure of a benzene compound with six methyl groups (carbon atoms with three hydrogen atoms attached) and reported the same results: a planar (flat) molecule with perfect symmetry. Now imagine attaching six more identical rings of carbon onto the perimeter of the first ring, replacing the six hydrogen previously occupied in benzene. Then add more identical rings onto that perimeter. Keep doing this forever. Eventually, you begin to fill out a honeycomb lattice of interconnected hexagons where every carbon is identical. Extended for hundreds or thousands of repeating units, benzene becomes graphene.

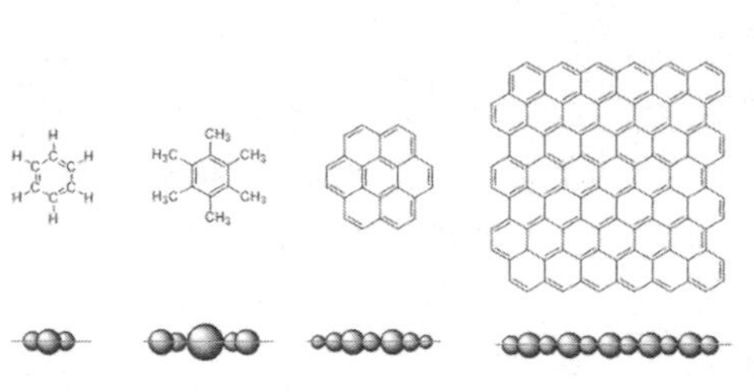

Figure 1-2: Top: Skeleton structures of different benzene structures eventually become graphene. Bottom: On-edge perspective drawings of carbon atoms, showing the flat structure of each molecule on top. (Image by Joseph Meany.)

If you look at the illustration above, the gray blobs represent a rough edge-on perspective view of the carbon atoms all lined up in a flat row. The hydrogen atoms have been eliminated from the structures for easy viewing. Benzene and the benzene with six methyl groups added are perfectly flat. The extended ring structure, commonly called coronene, is also a flat structure. Coronene falls under a group of molecules called *polycyclic aromatic hydrocarbons*, or PAHs for short. As their name suggests, PAHs are molecules that contain many connected rings (polycyclic) made from carbon and hydrogen (hydrocarbons), where the rings are chemically related to benzene (aromatic). Extend that out to a great distance, with all of these carbon atoms interlacing with one another in a molecular "chicken wire" structure, and a single graphite layer is perfectly flat as well. Built out, ring by ring, PAHs would eventually "become" graphene, although it is unclear to chemists at the moment at what point PAH properties would be indistinguishable from graphene. Current estimates predict that several hundred repeating ring units might be required before a PAH could be considered "graphene," although the number may as easily require thousands of tessellated hexagonal rings before it is chemically graphene.

In 1924, two groups of researchers independently reported that the crystal structure of graphite is a hexagonal net of carbon atoms arranged in flat pancaked layers. They showed, through analysis of small crystals of graphite, that each layer is stacked one on top of another, kind of like the mineral mica. If you have ever encountered natural mica on a hike, you know that it is fairly easy to peel off a single sheet. This single sheet is exceptionally thin and almost perfectly clear. You can bend it, and it weighs next to nothing. This is a perfect analogy for graphene—if one could somehow peel a single layer from the graphite structure then it would be almost entirely transparent, incredibly flexible, and extremely light.

Recall that graphite and things like coal and diamond are each a different type of carbon. But how, if all atoms of an element are identical, can that be true? It comes back to the idea of bonding and how atoms share electrons with one another. Certain elements can connect atoms in different ways that give distinct properties to each form. These forms, called allotropes, can produce wildly different variants in properties, depending on the exact configurations of the molecules.

For example, sulfur has many interesting forms. It can be a colorless gas as two sulfur atoms (S_2) or a bright red gas where three sulfur atoms bond together as S_3. It has several solid forms, of which the bright yellow S_8 can be mined as large volcanic crystals or as a yellow powder that's mixed in with charcoal and other ingredients to make black powder for fireworks. At high temperatures and pressures, sulfur forms solids that can conduct just like metals do. Phosphorus, carbon, oxygen, and many other elements each have different ways in which the atoms can attach and that affect their properties. We're all familiar with the clear, colorless oxygen gas that we breathe in order to live. But if you take that oxygen and smash it really, really hard—using almost a million atmospheres worth of pressure—then it will solidify and turn a deep red color.[19]

The concept that elements can change their forms and bonding structures when put into extreme environments is familiar to many people already, especially miners. As prehistoric trees and swamp plants died and were buried, these carbon-containing materials were put under increasing pressures and temperatures. Over time, the carbon atoms were pushed together more and more tightly. Other elements reacted with the surroundings and were pushed out. Water, hydrogen sulfide, and other lighter molecules were pushed away, concentrating all the carbon together. As time passed, these reactions kept happening, and the carbon atoms squeezed tighter and tighter together. Eventually, the impurities were all pushed away and left a seam of coal—amorphous carbon. But if this seam is kept underground for even longer, and squeezed harder at hotter temperatures, then the carbon atoms will start to rearrange themselves. These carbon atoms will start to form bonds in flat planes and begin to sandwich one another. Anthracite coal, the highest grade of coal, under high temperature and pressure will undergo metamorphosis into graphite. From chaos, order emerges. Eventually, that graphite is mined from the Earth and put into pencils or into bearings as a lubricant or it can be incorporated into high-tech applications as will be explained in later chapters.

Amorphous anthracite is one allotrope of carbon, while graphite and diamond are others. We spoke about a few different allotropes of carbon earlier.

Charcoal is one well-known form. Logically, it smells burnt, the concentrated bitter scent of an extinguished fire. Running your finger along the wood's scorched grain, it feels smooth. Running crossways to the grain, the charcoal is rough and frictious, which leaves a crumbly black residue on your finger. It will crush into a fine powder with little to no trouble. When it's mixed with sulfur and potassium nitrate, you get gunpowder. It is one of the oldest known pure elements, even if people didn't know it at the time. Charcoal has been known since humans discovered fire and has been a critical resource since the dawn of smelting.

Diamond has a different role to play than charcoal in society. The tactile and olfactory nature of diamonds is unremarkable; it is the optical clarity and refraction properties that excite consumers' interests. Diamond's hardness makes it an industrially critical material in saws, sandpapers, and other high-stress applications. In an 1814 experiment that would probably cause a few gemologists to choke on their morning coffee, Sir Humphry Davy traveled to Florence, Italy, and was able to procure a high-quality sample of diamond. He put the diamond in a bell jar with a pure oxygen atmosphere, and, using a lens to focus the rays of the sun as an adolescent might do with a magnifying glass, he caused the diamond to catch fire. "The light it affords is steady, and of so brilliant a red, as to be visible in the brightest sunshine," he remarked in the *Philosophical Transactions of the Royal Society of London*.[20] When the diamond had finished burning, no residue was left behind. No ash, no strange metal oxide powder, nothing. Instead, Davy showed that the resulting gas was pure carbon dioxide. He produced an identical result from burning a piece of charcoal in the same apparatus. He concluded, then, that diamond and charcoal must have the same composition. They must be made from the same exact stuff.

Another crystalline form of carbon is graphite. It is the multilayered form of graphene you picked up this book to learn about. If you grasp a lump, it is smooth to the touch and slick almost to the point of feeling oily. Repeat this many times, and you will notice a faint gray buildup on your finger. If you drag the lump across a piece of paper, you'll notice that it does not crumble as coal will. Rather, it will slough off tiny little flakes in a gray line that you'll recognize as pencil marks. It wasn't until this century that the isolation of single graphite layers was recognized, and the later chapters of the book will delve deeply into this topic.

The baby of carbon allotropes are the *buckminsterfullerenes*, usually abbreviated to fullerenes, or popularly known as buckyballs. The "ball" designation comes from the molecule's shape—fullerenes are hollow spheroids of pure carbon. The original buckminsterfullerene is C_{60}, the best known of the fullerene class. Many other sizes of fullerene cages exist as well. They are the most recently discovered class of carbon allotropes, originally made from lasers blasting away chunks of graphite in a vacuum chamber. Later, scientists discovered how to make them from electric arcs, which we will talk more about in later chapters. Interestingly, fullerenes can be isolated from the black sooty material produced by candles, torches, and lamps. The next time you light a candle, hold a glass or a plate just above the flame. Do you see that buildup? You've just made some fullerenes! This substance, called lamp black, has been used in inks, cosmetics, and colorants since ancient times. In fact, lamp black is black because the different sizes of carbon clusters and the other byproducts in it absorb all the visible light. Each particle in that soot absorbs a unique color (wavelength) of light. If you were to dissolve pure fullerene samples of into different jars of benzene, then you would have quite the rainbow of samples as each fullerene would color the solution according to the wavelength of light it absorbs. Without getting too much into the weeds, each fullerene size absorbs colors differently due to the size-dependent electron orbitals resonating more or less efficiently with different wavelengths of visible light. This would directly translate into what your eyes would see as purple, orange, or yellow. Babylonians, Egyptians, and other cultures used this high-tech material to darken their eyes and lashes—imagine the implications this could have for retrofuturistic fiction. Fullerenes hide in plain sight as a part of typical soot powder and, like many great discoveries, were observed by accident.

Related to the buckminsterfullerenes are carbon nanotubes, which are sheets of graphene that are rolled up on themselves like cardboard wrapping paper tubes. They are usually capped on each end with hemispherical structures, basically half a buckyball on each end. Since they are generally much longer than they are wide, even being up to a million times longer than wide, carbon nanotubes are sometimes considered to be single dimensional materials. Their threadlike and wirelike properties offer considerable opportunities for making new structural materials and composites that are also electrically conducting.

But how do we possibly know that these different forms are, in fact, different? Is there a machine that tells us what these molecules look like? Can we magically take pictures of molecules to study as a biologist would take a picture of some small creature? Absolutely. And, of course, it is *science* and not magic. Crystals of molecules or atoms are routinely measured by a process called "crystal x-ray diffraction," where high-energy beams of x-rays are bounced off the electron clouds of the crystals. The bouncing of the beam occurs in a predictable and repeatable way, which is mathematically unique for each crystal type. From the interference patterns created by x-rays overlapping and changing their amplitude, crystallographers can determine with very high accuracy where atoms in a molecule go, which determines their shape.

In the early 1900s, this was a new and exciting field in which to work, and the mathematical theory behind x-ray crystallography was initially developed by Max von Laue.[21] X-ray crystallography was so groundbreaking that entire teams of physicists, chemists, and geologists clamored to examine samples of mineral crystals or organic crystals in a way that was never before conceived. Little wonder, then, that the discovery of diffraction by crystals earned the 1914 Nobel Prize in Physics for von Laue, followed immediately the next year by the 1915 Nobel Prize in Physics being granted to a father–son pair, W. H. Bragg and W. L. Bragg.[22]

The Braggs observed that crystalline organic (i.e., carbon-based) molecules could disperse x-rays in patterns that were characteristic of the molecules being analyzed. In other words, firing x-rays at carbon-based crystals would let the Braggs "see" the molecules in which they were interested.

This was not an easy field to enter. One had to understand very complicated mathematics in order to be able to decipher the hidden meaning behind otherwise useless bright spots and dark pits on photographic plates. In the days before automated computing, people who analyzed crystal x-ray diffraction data slaved for months over the calculations, turning out a few new analyses per year. It was a long and arduous practice, and if the data you recorded was inaccurate, you could potentially waste months on a dead end before retaking better data. The early literature on crystal analysis is fraught with examples of researchers calling each other out over misanalyzing some seemingly minor detail that invalidated whole interpretations of crystal structures. Since the introduction of computers into the field of x-ray crystallography in the 1960s, crystal structures have been easier to solve on a more frequent basis. Nowadays, the data collection for crystals can be done overnight and analysis of the data can be completed in a few days. This led to incredible advances, especially in medicine, where crystals of proteins may be used to help figure out the shape and composition of drugs that will best target specific maladies.

A protégé of W. H. Bragg, Mme. Kathleen (Yardley) Lonsdale was a remarkable person. She was born in Ireland in 1903 but raised in Britain due to hardships that her father experienced while she was young. She was a pioneer in her own right within the field of x-ray crystallography, eventually becoming the first female president of the International Union of Crystallography. During her primary schooling, she had such an insatiable appetite for studying the natural world that she left her original high school (Ilford County High School for Girls) for the boys' school, because the girls' school did not offer courses in the natural studies. She soon graduated from high school and entered the Bedford College for Women at age sixteen. There, she excelled and maintained numerous scholarships. Her high aptitude did not go unnoticed, and soon the Nobel laureate W. H. Bragg recruited her to work in his lab. Through this work, Lonsdale went on to have a remarkably successful career thereafter. In 1945, she became one of the first woman elected as a Fellow of the Royal Society, the UK's version of a national academy of science. (Another woman was elected the same year as she, a microbiologist named Marjory Stephenson.) Shortly before Lonsdale's passing in 1971, a new form of diamond was found in meteors. This new mineral was dubbed lonsdaleite in her honor.

Interestingly, only one woman prior to Lonsdale had been nominated for a Fellow position with the Royal Society. Hertha Ayrton was nominated in 1902, the year before Lonsdale was born. However, the Royal Society dismissed her application based on her being a woman. Ayrton was nonetheless a prolific researcher and mathematician, and, in 2010, she received posthumous recognition as one of the top ten most influential female scientists in Britain's history. Lonsdale was also on the list.[23]

While collaborating with W. H. Bragg, Lonsdale collected samples of graphite and was able to use x-ray diffraction to determine its structure. By this time, it was well known that graphite was another interesting type of coal, and so therefore had the molecular formula of C, pure carbon. X-ray diffraction was a tool that would finally elucidate why diamond and graphite look so different from one another yet still have the same fundamental building block. What Lonsdale found was curious but not earth-shattering at the time. Hexagons of carbon atoms extending out in flat sheets created stacks with each other. While the atoms in the same sheet were relatively close to one another, the distance between atoms in different sheets was much larger. This directly implies a great discrepancy in how strongly in-plane and out-of-plane atoms interact with one another. In other words, this means that the atoms are more strongly bonded to others in the same sheet than the sheets are bonded to each other. From that time, for nearly eighty years, a question remained to puzzle researchers: *Could one isolate just one of these atomic sheets?* What kind of properties would just one sheet have? Monolayer graphite has become known as *graphene*, and we are now on the upswing of the Graphene Revolution.

From the crystal structure of graphene, certain speculations can be made. The carbon bond lengths within a sheet (carbon-to-carbon) suggest that graphene is *aromatic*, meaning here that the atoms are strongly bound to one another in delocalized clouds. If that is true, then graphene should be a pretty good conductor. This made sense to scientists of the time, since graphitic carbon rods had been used as electrodes for various manufacturing processes for nearly half a century by the 1920s. In fact, you can even try this yourself at home. Take a pencil, cut away the eraser, and sharpen both ends. If you connect a multimeter or voltage tester across the pencil, you can measure the inherent electrical properties of that *particular* pencil. You can even make a functioning graphite circuit on a piece of paper simply by drawing dark lines on the paper with a pencil and connecting a battery. If you attach a light-emitting diode (LED) to the circuit, the LED will light up!

Other properties of graphene were not so well assumed during the last century. Graphite may be opaque and gray, but what would a layer of a single atom look like? What other properties lurk, just waiting to be unearthed?

Graphene's flat structure is important to the way that it acts as such a strong, flexible, and conductive material. The strong bonds between atoms in the flakes but weak attachment between layers of flakes are what give graphite its lubricant properties. The layers can slip and slide past one another with great ease. Since the electrons are generally stuck onto the layer to which they are attached, this leads to the great electrical conductivity of the material in the plane but poor conductivity in the direction between layers.

If that doesn't immediately make sense, think of it this way. If you were an electron on a flake of graphene, you could move back and forth or left and right as you pleased. It would be, in an analogous sense, just like walking around on the ground. You can run around on a level and open field without impediment. The field opens four principal directions in which to move.

Moving up and down, on the other hand, is much more difficult.

Picture yourself on a field. It is a nice day out, and the grass on the field stretches out before you. Up in the sky, no clouds float around. It is a clear day, but above you are floating platforms of fields identical to the one you stand on. Imagine a world where we had dog parks floating in the air at different levels. When you're standing on the surface of a park, you know that all the surfaces are identical—you and your dog run freely around to play fetch and throw Frisbees. If you walk for a while, you'll reach an edge of your field with a sheer drop. Below, there are more identical fields. Other people walk and run around on the fields you're looking down on. You can see as you look over the edge that there are ladders that connect the different fields. Getting from one level to another, though, is going to be much more difficult. Facing the prospect of having to pick up and carry your dog to a new park, you are much less likely to switch layers, so you will continue to move around on the one upon which you happen to currently reside. You are much less likely, from a quantum mechanical standpoint, to approach a ladder and climb it. Changing the mode of movement, from running around to climbing, is energy and concentration intensive.

We have this material, then, which was initially an intellectual curiosity without a great deal of importance. Graphene originally didn't attract much interest from the scientific or business pursuits, as the graph below shows. Until the 1990s, the idea of graphene was seldom mentioned in the scientific literature; a couple of sparse references here and there every few years were all that appeared between 1900 and 1990. After carbon nanotechnology took off in the late 1980s, however, coinciding with the discovery of the fullerenes and nanotubes, interest rekindled in graphene. New analysis techniques, like scanning tunneling microscopy, allowed unprecedented resolution in picturing chemical systems at the atomic level. Dozens of papers began to be published every year, imagining how to isolate and characterize this elusive material. It wasn't until 2001, when Novoselov and Geim isolated graphene using simple Scotch Tape that graphene research really began to enter mainstream consciousness. The whole history of this discovery is the topic of chapter 3.

Careful inspection of the graph also shows a sharp increase in publication rate at exactly 2010. This is the year that Novoselov and Geim were awarded the Nobel Prize in Chemistry for their discovery nine years earlier. In the six years since that award, hundreds of thousands of papers were published and billions of dollars invested across the globe in this wondrous material. As you'll see in successive chapters, though, the path forward is not without its own darker corners.

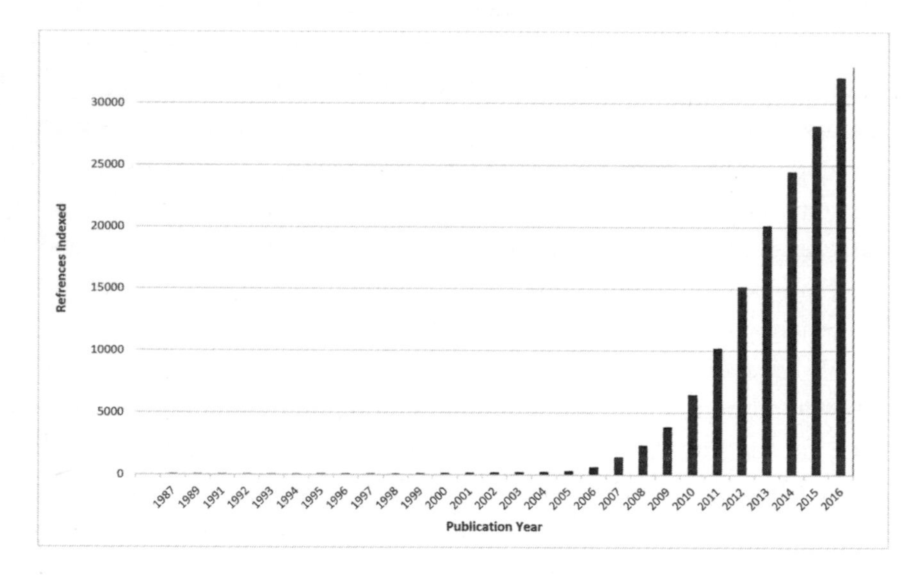

Figure 1-3: Google Scholar citations by year. (Image by Joseph Meany.)

Chapter 2

WHAT HAPPENED TO THE OTHER CARBON MIRACLE MATERIALS?

Articles within the Science and Technology sections of newspapers these days glow with anticipation of the marvels graphene could deliver. "Wonder Material Could Harvest Energy from Thin Air," exclaims one CNN article. The *Washington Post* suggests, "Why You Should Take Note of Graphene." "Bend It, Charge It, Dunk It: Graphene, the Material of Tomorrow," promises an article from the *New York Times*. A *Time* magazine article simply states, "Graphene: The Material of Tomorrow."

"Tomorrow" is a common thread among these articles. The futuristic applications of graphene are reported with a sense of wonder and the feeling that something big is lurking just around the next corner. The idea of graphene sparks excitement and a little bit of tension mixed with hope and hype. It inspires the same kind of bright optimism and futuristic dreaming as the Disney projects Tomorrowland park and the (canceled) Experimental Prototype Community of Tomorrow—better known as EPCOT. Walt Disney was certainly a futurist thinker, and he was inspired by the promises of "tomorrow" as a guide to his vision. In fact, one of his inspirations for the initial designs behind his EPCOT (which was distinct from the Epcot theme park in Walt Disney World today) was a book called *Garden Cities of Tomorrow*. But "tomorrow" was yesterday for many developments already out of the laboratory. Mobile phones and video chat used to be imaginations of science fiction. Rockets, wireless charging, and robots were "tomorrow" for the futurists of the past. They have matured in their own times, and now it is our turn to imagine tomorrow. Buckminster Fuller, futurist and architect whose design of the geodesic dome was adopted for EPCOT, released his own vision of a promising tomorrow in his 1938 book *Nine Chains to the Moon*.

There are many examples of disruptive inventions touted as "the next big thing that will change the world." We will explore a number of these in chapter 8, some of which were fantastic successes while others were spectacular failures. "Tomorrow" came for some of these inventions and never arrived for others. For example, we do have the internet but we don't yet have cold fusion. These are broad examples, though, that helped shape the technology we use (or don't use) today. But what about the other materials made from carbon, fullerenes and nanotubes, the materials that promised upsets in the fields of medicine, electronics, leisure, and art? Fullerenes had their day in the 1980s, and excitement around nanotubes peaked in the late 1990s. Beginning in earnest with the discovery of fullerenes in 1985, scientific editorial pieces have frequently exalted the nearly endless possibilities of high-tech materials made from carbon to transform everything from automobiles and architecture all the way through to cooking and clothing. The editors of the journal *Carbon*, in a 2016 editorial, summarized the many remaining challenges that face those researching graphene and graphene-related materials.[1] They outlined the challenges they perceived and described what they would like to see researchers in the field address. They encouraged researchers to submit articles for possible publication in the journal. Carbon (the element) has been a significant driving force within nanomaterial research since the beginning of this century, with many different applications in mind. One of the significant challenges facing carbon is extending its two-dimensional properties into three dimensions. The authors of this editorial also cover how so-called zero-dimensional fullerenes alongside one-dimensional nanoribbons and carbon nanotubes will fit into a scheme to create an entire tapestry of related carbon-based specialty materials. These chemicals will be guided both by the shape of the structure formed and the type of bond(s) dominating the carbon-carbon interactions.

Before we get to tomorrow, however, we should talk about yesterday for context. It is certainly helpful to understand where our understanding of carbon-based nanomaterials came from. The explanation begins, appropriately enough, with none other than the "Queen of Carbon," the late Mildred Dresselhaus.

Professor Dresselhaus, or "Millie," as she was commonly known to those close to her, came from a poor upbringing in the New York City borough of Bronx. Her schooling options were initially limited, but she was very personally driven and was able to find scholarship opportunities not afforded to her classmates. She studied hard and entered Hunter College High School for Girls at the age of thirteen. She encountered a physics teacher at this school, Rosalyn Yalow, who mentored Dresselhaus and encouraged her to pursue a career within the sciences despite societal pressures to enter more "traditional" professions for women at that time.[2] Yalow would go on to win the 1970s Nobel Prize in Physiology or Medicine for her work on radioimmunoassays, a technique to measure certain types of molecules within the body.[3]

Dresselhaus graduated from Hunter College and completed her graduate studies at the University of Chicago. Following her time in Chicago, she moved with her new husband, Gene Dresselhaus, to Cornell University, where he found work as a professor. Millie Dresselhaus was able to became a post-doctoral researcher (postdoc, for short) at Cornell, working on the physics underlying superconducting materials. Postdocs have always been temporary positions, by design, so Millie had to keep an eye out for possible new directions for research at another institution, where she might find a permanent career.

The couple began to search for a permanent home for themselves, but it was rare at that time for an institution to hire both married scientists. Today, hiring two scientists at once is much more common, but still it poses a challenge for married professionals—especially academics. The issue is often referred to as the "two-body problem," a wink and a nod to the concept in Newtonian Mechanics by the same name. Married graduate students and postdocs must take into careful consideration their own career trajectory and goals, hoping that their spouse can also find gainful work at a new university or lab. If the married couple perform research in the same field or one closely related, then the problem is compounded, as institutions do not often have two openings in the same department at the same time.

As Mildred Dresselhaus was searching for directions in which to launch her independent career, she reached out to several mentors for advice on both where to locate and in what research area to direct herself. Her work at the University of Chicago had focused on creating high-temperature superconductors out of the element bismuth (you know bismuth as the colorant and a component of the active ingredient in a certain bright pink stomach medicine), but Dresselhaus knew that high-temperature superconductors were beginning to fall out of scientific vogue.[4] Fortunately, she and Gene were both invited to MIT's Lincoln Laboratory by Dr. Ben Lax in 1960. It was there that she began her efforts to understand the underlying physics of how *charge carriers* move through semiconductors.

Charge carriers consist of electrons and, unintuitively, something called a "hole." A hole is simply the lack of an electron where one could or should exist in an atom. The absent negative charge creates a hole, or empty spot, that behaves like a positive charge. Semiconductors, like silicon and germanium, are elements used in computers and modern electronics. The Dresselhauses analyzed many different types of materials, and when they decided that they needed to switch gears, Gene suggested that Millie move onto looking at characterizing graphite-like materials.

In 1960, graphite and diamond were the only two known allotropes of carbon. The hexagonal stacks of carbon molecules in soft graphite were distinct from the hard cubic edges of those in diamond. Fullerenes and nanotubes were still some decades away from discovery, and even lonsdaleite would not be found until much later that decade. Carbon fibers, which were discovered in 1957, were merely an expensive academic curiosity rather than a new blockbuster material. Since carbon is so common, and graphite is dug out raw from the ground, it did not attract much attention. Graphite continued to fill the same primary niche that it had since the middle of the 1500s and one with which you are doubtlessly still familiar—pencil lead.[5] It was so generally disregarded by the scientific literature during the mid-twentieth century that in interviews before her passing in February 2017 Dresselhaus recalled, "There were three papers per year in the world, and I think they were almost all mine."[6] She went on to describe how her work on "boring" graphitic materials, combined with the pervasive air of sexism in science, allowed her the flexibility necessary to raise a growing family. Nevertheless, she persisted.

Millie, Gene, and Dr. Lax all published papers on graphite, using lasers and magnets to figure out where the electrons are energetically located within a graphite sample. In the scientific jargon, they determined its *band structure*, finding a lot of interesting curves derived from complicated mathematics and experimental data. Just prior to the trio beginning their experiments, a colleague used graphene (which was known as two-dimensional graphite at this time) to calculate how electrons were arranged in a single sheet of the graphite. The calculation for a realistic three-dimensional structure was greatly simplified by reducing the problem to a planar structure. This simplification is an assumption that mathematicians and physicists often make in order to reduce impossible or extremely difficult calculations into more manageable pieces. Simplifications like this can sometimes lead to exuberant discussion in scientific journals and at scientific conferences. More than a few harsh words have been uttered over professional differences of opinion. People will defend and decry these simplifications all in the name of sound scientific advancement.

While it is true that simplifications do not properly encapsulate every detail of a complicated situation, the use of shortcuts makes many complicated problems available to answer. Those who took physics during high school or college might remember a version of this problem-solving tactic. When considering a complicated question, certain assumptions about the system can help make it more approachable. For example, one could treat many simultaneous forces on a ball as one larger force pushing on a single point, since a true and complete representation of the ball would be nearly impossible to represent with pen and paper.

The realm of experiment is never so neatly treated, nor so easily reduced. Despite graphene's high inherent crystal symmetry, the flake sizes are small and irregularly shaped. A hunk of graphite dug out of the ground that you can hold in your hand is never crystalline across the whole piece. It is instead made up of many small crystals all mashed together. Scientists who study crystals like this call bulk graphite *polycrystalline*. Drop that word at your next wine party, and I'm sure you'll get some polite nods. High crystallinity is necessary to measure the band structure so that the electrons are able to travel only along and between parallel stacked graphene sheets. If the individual flake planes are tilted or bent, they cause a disturbance, the signal gets muddied, and researchers can't make conclusive observations. Think back to the infinite floating planes in the last chapter. If the planes crisscrossed, the dogs could move between the layers at will, and keeping them in check would be impossible.

In the 1960s, companies were not exactly battling one another to prove that they could produce the highest quality single-crystalline graphite. Mostly, those companies that did produce graphite focused their attention on making pencils in the same way that they had for centuries. There was no way for Millie Dresselhaus to simply call up a supplier and say, "I would like your most highly single-crystalline graphite, please." She could not simply hop on the public train down to a corner market for some high-quality C. In order to do the experiments correctly, Millie, her husband, and Lax had to find a more appropriate material. In 1960, Millie discovered that high enough temperatures and pressures could form diamond. This helped, in part, to inspire researchers to explore extreme conditions with pure carbon to see what interesting and unusual things could happen.

Shortly thereafter, in 1962, two scientists, L. C. F. Blackman and Alfred Ubbelohde, heated methane in a chamber and came up with an interesting and unusual result. After careful measurements, they found that hydrogen had separated from the methane carbon, and a residue was left behind. They found the residue was graphitic in nature. It consisted of interlocked hexagons of carbon extending in a plane, and the planes stacked on top of one another. The crystals could even be grown larger than any samples one would find in mined graphite. Due to the great degree of crystallinity in the material, the Ubbelohde group is credited with creating the very first *Highly Oriented Pyrolytic Graphite*, or HOPG. Ubbelohde and Dresselhaus entered a collaboration together, and it was from his samples that Dresselhaus successfully deduced the electronic nature of graphite.

This result was a special application of a technique known as *Chemical Vapor Deposition*, or CVD for short. CVD is used for many different applications, and it is not merely limited to methane or other carbon-based gases. CVD of silicon atoms or even more complicated molecules is possible with the right chamber and proper conditions. It all depends on what you want to make. For example, the right conditions in a CVD chamber won't form graphite but will instead form diamonds. Companies exist that will turn your loved ones' ashes into diamonds. Talk about diamonds being forever.

In 2016, British chemists figured out how to use CVD to recycle and use spent nuclear material in a clever way. First, they heated graphite that had been used to coat and shield radioactive cores into a gas. This graphite was enriched as a heavy form of carbon that had absorbed loose neutrons from the uranium fission process. This heavy carbon is radioactive (it has a half-life of over five thousand years), and so it would produce a low amount of power for a long time if we were only able to capture that energy. But how could we do this safely (nobody wants to cozy up to a nuclear battery) and easily? The clever solution came through CVD. The researchers sublimed carbon-14 (^{14}C) enriched graphite and formed a diamond out of the heavy-carbon-enriched gas. To protect users from excessive high-energy radiation, the small diamond was coated with diamond made from non-radioactive carbon. The power would come when the heavy carbon decayed. The ^{14}C nucleus would emit a hot, high-energy electron and another particle, called an antineutrino, transmuting the ^{14}C into a nitrogen atom, ^{14}N. The electron that was emitted can now move about in a circuit, which allows us to create a device that can tell time, take pictures, or calculate something. The possibilities are only limited to low-power applications that would be needed for a long time.

When CVD was used to make HOPG, and samples were finally available to measure graphene's properties on the nanoscale, working out the properties of graphene on an *experimental* scale (aptly named, something on the experimental scale lends itself to manipulation and measurement in the laboratory) finally began to appear possible. As it was, this work did not immediately yield isolated graphene sheets. However, Dresselhaus and her students worked out the finer details with respect to how electrons and phonons (vibrational waves) move about in graphene compounds. This work led them to investigate *intercalation compounds* of graphite. Intercalation compounds are interesting materials. They are formed when a material like graphite, called a *host*, accepts and mixes with a separate material, called a *guest*, to form a completely new three-dimensional structure. The weak interaction between the layers within a host graphite sheet allows other guest atoms or molecules to slip between them.

For example, potassium metal can react with graphite to form an intercalation mixture. When the potassium is melted and poured onto the graphite, the atoms of potassium push their way in between the layers of graphite and nest within the hollows created by the carbon rings. The potassium starts out as a shiny silver metal; powdering the graphite turns it into a black powder. Interestingly, when the two are combined and the intercalation is complete, the resulting powder is a deep bronze. Just as amazingly, this combination of potassium and carbon were even found to achieve superconductivity,[7] although at too low a temperature to be useful for widespread adoption. But what is happening here? Why would the potassium even "want" to mix with graphite in the first place?

In this particular case, the potassium metal is neutrally charged. It has one electron in the outer shell that it is just dying to be shed. If you've ever had the chance to throw potassium or sodium into water, then you will have undoubtedly noticed the ensuing reaction in all of its bright, loud, and energetic beauty. Both metals are extremely reactive with water, and this is because of this one extra electron. Graphite, though, is extremely electrically conductive. One of the benefits underlying this property is that graphite can spread out those extra electrons, smearing the repulsive forces over a wider area to stabilize the potassium ions within the resulting material. The potassium becomes a guest to graphite's host. Since one is adding extra electrons to the system with the added potassium, the electronic symmetry of the resulting mixture is different from plain graphite as well. However, graphite does not need to only be an electron acceptor from donor atoms or molecules. Graphite itself can act as a donor to appropriately strong acceptors. Just as an extra electron's negative charge may be smeared over the graphite sheet, the resulting positive charge from an electron's removal (the hole mentioned previously) may be stabilized in the same manner. Semiconductor physicists don't always like the term "remove an electron," so if you ever find yourself in conversation with one, make sure to sound knowledgeable by saying "injects a hole," instead. The charge balance accounting still works out the same.

The broad array of problems (air unstable, explosive, difficult to work with) encountered within intercalation compounds presented interesting lines of research for a condensed matter physicist like Mildred Dresselhaus. Studies on intercalation compounds kept her occupied through the 1970s, and into even the beginning of the 1980s. In the 1997 edition of the *Annual Review of Materials Science*, she wrote,

> For many years, especially early in my career, carbon science was a backwater field, considered by many to be too complicated and by others too mundane. . . . My students, coworkers, and I also enjoyed working in a field that was not in the limelight, where one could do careful work and take the time to understand what was going on.[8]

And take her time she did. Dresselhaus worked with the Lincoln Lab exclusively for eight years, eventually becoming the first female full professor at MIT in 1968. She transitioned part-time into teaching, which she took very seriously. Eventually, she left Lincoln Lab altogether to concentrate on her duties at MIT.[9] Her husband, meanwhile, continued to have a very productive career at Lincoln Lab. Together, they pioneered new work with carbon, and we will see more of their contributions scattered throughout the rest of this chapter and book.

The early history of carbon chemistry involved a lot of dirty work. It was the kind of work where, at the end of the day, you would be covered in soot and char, your saliva and mucus would be discolored, and anything not covered by your smock or apron would be blackened in the same way. Shaking the coal dust from your hair, you would wipe thick gray sweat from your brow, darkened from the powders suspended in the air. This doesn't do any good, of course. Rather, it just smears what feels like gritty grease across your face. The workshop's sink is stained from long months of constant abuse in these environments.

Early carbon chemistry focused on understanding the mysteries behind the most flamboyant of Aristotelian elements, fire. Experiments focused on combustion and respiration, two processes that were chemically linked in early scientists' minds, although they were not clear how. Early researchers discovered that burning coal generated carbon dioxide, a gas that did not support respiration. Candle flames and rodents placed in bell jars filled with the gas expired. Carbon dioxide is also colorless, flavorless, and odorless, making it ever more mysterious. Exhalations of this gas into a solution of calcium hydroxide[10] turn the solution from colorless to a milky white suspension. Carbon dioxide in water turns into carbonate, which then precipitates out as a fine powder of calcium carbonate.[11] This was before the days of proper chemical nomenclature, so depending on where the researcher was from a newly discovered chemical gained a different name (or several different names, since standardization didn't exist yet). Likewise, as the gas was notoriously unreactive, its exact chemical composition remained unknown for some centuries. This was puzzling; how could coal and people produce the same gas when no anatomical investigation could point to our own internal combustion engines? In other words, where was life's fire?

Today, this may seem to be a laughably obvious question to answer. With the benefit of hindsight and six centuries of the world's best minds pouring over this question, we now understand that carbon dioxide is a waste product of our cellular energy production. Our internal combustion engines are organelles called *mitochondria.* In the days of alchemy and early chemistry, the phlogiston theory[12] dominated scientific inquiry, only to be replaced by the modern understanding of combustion after chemists finally learned to balance chemical reactions—quantifying the idea of "what goes in must come out." From there, coal seemed to have become uninteresting; it had been tamed by the whip of science.

Graphite itself found two early uses, the first being as a writing tool in pencils. Due to this, Abraham G. Werner coined the term "graphite" in 1789, meaning "writing stone." Graphite's lubricating property was also used in lining cannonball casting molds for easier mold release, which allowed the British to increase their production rate. With added supplies of cannonballs, the British found graphite useful in naval warfare during the late 1500s.

Eventually, the discovery and development of the electric circuit spurred a cavalcade of research within physics and chemistry. Experiments and demonstrations across the globe created a flurry of letters and journal entries. Patents for inventions swelled within trade offices, as researchers scrambled to mark their intellectual property and to profit from marketing their discoveries. In 1800, Alessandro Volta invented the *electric pile*, the first chemical battery, allowing for the creation and storage of a chemical-electrical potential.[13] This early battery would create a constant source of controlled energy to use in experiments—a good and useful thing.

Almost immediately, someone thought to use electricity to create light. John G. Children, a friend of Sir Humphry Davy, demonstrated incandescence in platinum and charcoal soon after Volta's invention became available. In 1802, he placed a strip of charcoal between two wires connected to a battery. This creates a circuit. Charcoal is a poor conductor, so its natural resistance caused it to heat up and eventually begin to glow first red, then orange, yellow, and white. The heat and light were extremely brilliant, but this early demonstration would not be suitable for home use. It was not done in vacuum, so the charcoal burned away over time. Children was also able to perform a similar demonstration with platinum wire, getting it white hot and fusing two lengths together.

The feverish pace of scientific discovery in relation to electricity would continue throughout the entirety of the 1800s. In 1808, Davy demonstrated an electric arc between two carbon terminals. In an experiment before the Royal Institution of Great Britain, he generated light arcing between two carbon rods separated by a gap of air. As you know, lightning is bright and loud. The crackle of this carbon-based electrical discharge was sure to impress onlookers as readily as Tesla coils or Van de Graaff generators do with static electricity today. In 1892, John Tyndall of the Royal Institution published a book titled *Fragments of Science*, which collected historical essays on the development in different areas of science. He writes about Davy's demonstration,

> Davy was enabled to construct a battery of two thousand pairs of plates, with which he afterward obtained calorific and luminous effects far transcending anything previously observed. The arc of flame between the carbon terminals was four inches long, and by its heat quartz, sapphire, magnesia, and lime were melted like wax in a candle-flame; while fragments of diamond and plumbago rapidly disappeared, as if reduced to vapor.[14]

"Plumbago," it should be mentioned, was graphite ore and not lead, as some astute viewers of the periodic table might notice. The chemical symbol for the element lead is Pb, after the Latin *plumbum*. The confusion persists today each time someone refers to pencil "lead." The confusion is understandable. In fact, the miners who discovered graphite in the hills of Great Britain are to blame for the mix-up. Unskilled in the alchemical arts, they thought the slick, gray, lustrous material was actual lead. They must not have had a lot of firsthand experience with actual lead, though, because the density difference alone would have been a dead giveaway. It makes sense, then, that graphite and diamond should both react to vapor, as they oxidize to carbon dioxide in the presence of air and heat. They are literally vaporizing through this chemical reaction sparked by the electric arc's plasma stream. In fact, Antoine Lavoisier proved that diamond and graphite have the same chemical composition in the late 1700s when he burned several carbon-based materials in an oxygen atmosphere and proved they all released the same product gas. While the concept of plasma was not understood in Davy's time, we now know that this arc of flame is a superheated stream of ionized matter—lightning on a lab scale.

Throughout the 1800s, arc lamps developed in complexity and utility, eventually achieving commercial success in industrial lighting and general use. Carbon electric arcs were not housed in protective vacuums as Edison's incandescent lights were, and, as such, were consumed over time. This limited their adaptability for home use, despite producing brilliant light. While some inventions increased the efficiency of carbon arcs over time, the long-lasting incandescent bulb proved to be better for consumers, and the arc lamps fell into niche uses. Incandescent lights eventually took over and formed the basis for our home electrical revolution in the early 1900s.

Eventually, many different independent inventors figured out that heating a filament in an inert atmosphere or vacuum improved the operation of electric light bulbs. Joseph Swan successfully developed an electric light bulb in the 1860s that was competitive enough for the commercial market. Swan's light bulb was constructed using graphite-covered paper as the lighting filaments. These carbon-covered filaments had appropriate electrical properties for making incandescent lightbulbs because their resistance was high enough to generate light from resistive heat but the conductivity was high enough that power generation was not a commercial obstacle. Swan competed with Thomas Edison for dominance of the electric lighting market through the 1870s, and eventually they became business partners. Over time, Edison's ability to use the press surrounding his partnerships and accompanying business interests led to him being credited with the invention of the modern lightbulb, and he is the one that we read about in history books.

Last chapter, we briefly introduced Hertha Ayrton, a British mathematician and physicist who was the first female member elected to the Institution of Electrical Engineers (IEE) for her research on electric arc lamps.[15] In fact, she became such a renowned expert in the field that in 1902 she published a book drawn from her own experiments and the review of others, titled *The Electric Arc*. Embarrassingly, due to her sex, she was not permitted to read a research paper before the British Royal Society in 1901. It was instead read by a male colleague, John Perry. In 1904, the Royal Society reversed its decision, and Ayrton was permitted to read a subsequent research article on wave-induced ripples within coastal sand. In 1906, she was awarded the Hughes Medal for her work. She continued working in physics and mathematics after her husband's death in 1908 until her own death in 1923. Her work cleared the way for truly innovative uses of the graphite electric arc, leading to huge advancements that resonated across metallurgy, carbon nanoscience, and engineering.

In 1958, a young and enthusiastic chemist named Roger Bacon[16] joined the staff at Union Carbide. He was tasked with melting graphite at high pressures and temperatures to find the elusive physical *triple point* of carbon. Although we used carbon in many important ways throughout history, the element (as a whole) was not entirely well characterized at this time. The triple point is a term for the temperature and pressure when the solid, liquid, and gas phases for a given material are all in equilibrium—*when they all exist together at once.* For example, the triple point of water would have a boiling beaker of water with ice floating on the top of the liquid caused by a unique interplay of the temperature and pressure. The same would be true for carbon: solid graphite would exist while flakes within the sample melt and slide around one another as a liquid, and evaporate away as gas. Thermodynamics is weird, we agree!

Careful experimentalists can operate machines that vary pressures and temperatures over huge ranges. These scientists can use tools to observe the phase of matter (solid, liquid, gas) that their sample adopts at any given temperature and pressure point. It is possible to draw a chart for most materials that maps their physical phases over a range of temperatures and pressures. These data points can help a scientist create a picture of where these phase transitions occur. Gasses, for example, will turn to liquid or even a solid with a high enough pressure (at a constant temperature) or a low enough temperature (at a constant pressure). This makes sense, because in the constant temperature case, molecules of gas are being pressed closer and closer together until they eventually have only very restricted mobility. It is then that they become a liquid. Once the pressure has increased again, to another, much higher, pressure, the molecules stop being mobile altogether, and the liquid freezes into a solid. In the other case, with constant pressure, lowering the temperature of molecules strips them of their *kinetic energy*. Kinetic energy is the energy that objects have when they move, whether they be a molecule, a bowling ball, or a planet. Removing kinetic energy results in the object slowing down. As the temperature goes down, eventually gas molecules get slow enough that their kinetic energy cannot overcome forces between molecules, and they coalesce into a liquid. As you know from water freezing, lowering the temperature of a liquid causes molecules to arrange themselves in a crystalline structure—forming solid ice.

As Bacon worked to determine the triple point for carbon, he was given tremendous creative freedom in how he would conduct his research. He used a setup very similar to the carbon arc electrodes described previously, but his apparatus was different in that it worked at higher pressures than the usual lamps. It was not long before Bacon witnessed something extremely interesting. During his test, he noticed that the graphite *sublimed* directly into a gas.

When Bacon switched his apparatus on, graphite sheets vaporized—sublimed—from the surface of the graphite block. They flew across the apparatus chamber and did something that nobody had ever recorded before. The gaseous graphite sheets, when the pressure within the chamber was below a certain threshold, coagulated into small solid rods. This process, called *deposition*—going from the gaseous to the solid state—was not the odd part. The odd part was in that these small solid rods formed. Imagine steam being released from a pressure cooker and instead of condensing above the oven, little needles of water grow on the range hood. Nobody had ever described such a weird phenomenon, and, fortunately, when he opened the chamber, the rods stayed intact.

In an interview with the American Chemical Society, Bacon recalled, "They were imbedded like straws in brick. They were up to an inch long, and they had amazing properties. They were only a tenth of the diameter of a human hair, but you could bend them and kink them and they weren't brittle. They were long filaments of perfect graphite."[17]

After careful analysis, Bacon confirmed what he had suspected—these were scrolls of graphite sheets stacked together side by side to create an elongated structure with high crystalline features. That's why he was so confident in calling it "perfect graphite." X-ray crystal diffraction studies helped him establish the crystallinity, but it was electron beams that helped him magnify these structures to see them from a new perspective. Part of his initial paper, published in the *Journal of Applied Physics* in 1960, demonstrated "the commonly observed fact that a decrease in diameter is accompanied by circumferential steps on the whisker and an increase in the transparency to the electron beam."[18] This means that the carbon fibers were built around a core in layers, like a paper towel roll around the cardboard center. Bacon was able to use the electron beam to evaporate segments of the carbon fiber. Graphite sheets would then flake off, exposing layer after layer of this tubular onion he was peeling.

This wasn't the end of Bacon's remarkable discoveries, however. Later in the paper he described an experiment where "a whisker whose outer layers were 'exploded' off by the passage of a heavy current through it" was put under the magnifier.[19] He described seeing a thin and hollow tube with the remnants of a few outer layers scattered around the rest of the image. This might not seem like a big deal, but it was a missed opportunity for Bacon to make another truly astounding discovery within this same set of experiments. He saw, but failed to recognize, the structures that would eventually go on to be dubbed carbon nanotubes. This was a fact he recognized later with a significant degree of humility, and the credit for recognizing carbon nanotubes for what they really are is a much more complicated process, which we will address shortly.

While Edison patented the idea of a hollow carbon-based tube within his early carbonized light bulb filaments, there was literally no way that he could prove such a hypothesis with the available technology at the time. Marc Monthioux and Valdimir Kuznetsov, in their 2006 editorial in *Carbon*, agree that Edison and Swan probably produced carbon nanotubes (though they were unrecognized as such) in their research.[20] H. P. Boehm even goes so far as to provide evidence that early experiments to produce silicon carbide by Edward Acheson in the 1890s formed synthetic graphite and carbon nanotubes as a byproduct of the high temperatures within the reactor.[21] While it is undoubtedly likely that Edison, Swan, and Acheson formed nanotubes, the first recorded photographic evidence for carbon nanotubes had to wait until the Transmission Electron Microscope (TEM) was invented. In 1952, two Russian scientists, L. V. Radushkevich and V. M. Lukyanovich, published TEM images as early proof of multiwalled nanotubes. Unfortunately, the paper was published in their native Russian during the height of the Cold War, so it went unnoticed and unread by Western scientists for several more decades.[22] Today, most science is published in English, although some journals do publish in local languages. These papers generally have a lower scientific impact as they are not sought after by nonnative readers.

The carbon fibers that we described above are related to carbon nanotubes in that the carbon fibers are usually built up around a hollow carbon nanotube core. They are distinct from multiwalled carbon nanotubes, though, because multiwalled carbon nanotubes have a constant and unbroken construction, whereas carbon fibers may be made from disjointed graphene platelets. Research on the growth of thin carbon fibers accelerated over the next two decades, until Morinobu Endo finally published evidence in 1976 that a single-walled carbon nanotube sat at the core of these fibers. While this was a remarkable finding, it was still not recognized as a watershed moment in science. The audience for the information, which was published in the *Journal of Crystal Growth*, was too narrowly focused to generate a great deal of excitement in the wider scientific community. Research continued through the seventies and eighties and was subsequently eclipsed in importance by the discovery of fullerenes.

The discovery of fullerenes did have one further unintended effect on carbon nanotube research, allowing scientists to accept the idea that carbon nanostructures could be hollow. Truly hollow spaces, with only the vacuum of the universe inside, were generally believed to be unstable over time periods familiar to humans. *Horror vacui*, "Nature abhors a vacuum." In our everyday experience, water or air permeates just about anything we can create. During the late eighties, scientists furiously debated the existence of carbon structures with hollow centers, and evidence built up to support the existence of such exotic materials. The idea that a molecule with a hollow center could exist was not easily accepted, but after it had been accepted, researchers came to another conclusion. Using fullerene molecules as endcaps for extended tube structures, by wrapping graphene around on itself, would create long, strong, fibrous molecules. Dangling bonds at the end of uncapped tubes would be unacceptable, as atoms at the tube termini would have unsatisfied octets and therefore be extremely reactive. The fullerene endcaps solve this dilemma by making sure all atoms have bonds in valid molecular geometries. These long, tubular molecules could be extremely conductive. It seemed that the stage was set for something new.

The discovery (as far as the wider world was concerned) came about in 1991 when *Nature* published Professor Sumio Iijima's "Helical Microtubules of Graphitic Carbon."[23] Carbon nanotubes had been discovered before, as we mentioned above, by two Russian scientists. While contemporary historians are earnest in their attempt to correct early editorial articles, which give Iijima full credit for the discovery, it is undeniable that Iijima's 1991 paper brought the focus of carbon nanotubes into scientific vogue around the world. Fullerene production had been improved by this point. Instead of being produced a few molecules at a time within the laser blasts of a vacuum chamber, fullerene molecules could now be produced in great quantities within an arc discharge.[24] The same contraption that gave us spotlights, lightbulbs, carbon fibers, and fullerenes now gave rise to another new form of carbon, the single-walled carbon nanotube. Edison and Swan, it seems, were right.

This time, however, carbon nanotubes did not continue in obscurity. They were no longer an interesting curiosity pushed to the corners of an esoteric science. This time, carbon nanotubes enjoyed the full attention of the wide scientific community, while riding on the waves of excitement afforded by the buckminsterfullerenes (which we talk about later in this chapter).

You may be aware of carbon fibers in high-end consumer goods (like bicycles and camping gear), where they lend a lightness and strength that other materials simply do not have. The industrial production of carbon fibers began in the 1960s and research into applications for this type of material ran parallel to graphite research and, after their discovery, to research into carbon nanotubes.

Commercialization efforts for carbon fibers have been more successful than carbon nanotubes to date, since they are cheaper to produce. Initially, carbon fibers were produced from the carbonization of rayon or other synthetic plastic fibers, although *polyacrylonitrile* (or PAN for short) is the current industry standard for fiber production. Eventually, if the benefits of carbon nanotubes in end-user devices can justify their higher cost, then we may see an increase in nanotube-derived composites. Until that benefit is clear, carbon fibers will maintain their markets. The electronic properties of carbon fibers remain attractive to engineers and as 3-D printing (also known as additive manufacturing) gains widespread adoption, then we may yet see objects printed with circuits embedded in their solid structures. Carbon-fiber printed circuits could see heavy-duty use in disposable electronics especially, but only after 3-D printing also adopts readily recyclable materials as their basis.

The end of World War II saw a dramatic shift in research into electronic circuitry, and in 1956 the Nobel Prize in Physics was awarded to William B. Shockley, John Bardeen and Walter H. Brattain "for their researches on semiconductors and their discovery of the transistor effect."[25] The development of the transistor in the 1950s did not go unnoticed around the world. As semiconductor technology was taking off, Professor Hiroo Inokuchi predicted that carbon-based molecules with distributed *p*-orbital electron clouds (such as benzene, naphthalene, anthracene, and, by extension, graphene) could someday be used as components in electronic circuits, replacing silicon. This idea was not immediately popular, but research in recent decades has shown the idea of molecular electronics to be extremely promising, especially for supplementing silicon devices. It was for his pioneering work on conjugated organic electronics that Inokuchi was awarded the 2007 Kyoto Prize. Traditional inorganic semiconductors may not entirely disappear, but it is likely we will see interesting hybrid devices emerge, as graphene and other carbon allotropes gain commercial support. We will briefly revisit the possible inorganic semiconductors developing in the near future in chapter 12.

It would be unfair to characterize molecular electronics as a field that focuses solely on carbon-based devices, though. While carbon is certainly an exciting focal point, elements like sulfur, selenium, gold, and iodine are beginning to find special niches for themselves as well. A great deal of work in synthetic organic chemistry exists, which allows chemists to create and modify molecules to the creator's specific desire. Instead, molecular electronics exists as a field of study to create functional devices based on properties we desire and can predict. From there, a blueprint, roadmap, or plan can be created to manufacture the components and finally assemble the device. A related analogy would be the creation of a skyscraper, or perhaps a dress.

A businessperson would approach an architect to design a building that fulfills his or her needs. The architect would take the design to an engineering firm to refine the vision and outline how to create the structure based on known principles. The engineering firm then would connect with contractors to make the individual parts and would hire a construction company to put the parts together. In the end, the building has gone from concept to concrete in well-defined steps. In the same way, a fashion designer would need to create a dress by understanding the occasion where it will be worn and selecting appropriate materials to accentuate the wearer's form. The creases and folds of clothing follow predictable rules, and applying those rules creates a fabulous finished product that started out as an idea or rough sketch.

At a conference in the 1950s, Colonel C. H. Lewis of the United States Air Force stated that

> We should synthesize, that is, tailor materials with predetermined electronic characteristics. . . . We could design and create materials to perform desired functions. . . . We call this more exact process of constructing materials with predetermined electrical characteristics Molecular Electronics.[26]

Richard Feynman, the bongo-playing Caltech physicist, gave a famous lecture on the fundamental ideas underlying nanotechnology called "There's Plenty of Room at the Bottom."[27] It is a remarkably accessible discussion on how to think about materials and the possibility of bottom-up material engineering (think 3-D printing, but on the atomic scale) rather than the top-down approach associated with chisels and saws to remove material. Feynman echoed Colonel Lewis as he talked about first considering what you want a material to do and then figuring out how to create it. He called for the exploitation of atomic manipulator machines to mimic or surpass the function of bulky materials, which would hopefully deliver on the promises of making circuits faster, smaller, and more efficient. Feynman studied the world of quantum physics and was awarded the Nobel Prize in Physics alongside Sin-Itiro Tomonaga and Julian Schwinger "for their fundamental work in quantum electrodynamics, with deep-ploughing consequences for the physics of elementary particles."[28] Part of their work in quantum electrodynamics (essentially, the field that deals with how light and matter affect one another on the atomic and subatomic scale) has since been applied to graphene and used to find interesting physical properties that weren't so much as dreamed of for this material before.

Until 1982, when the Scanning Tunneling Microscope was developed by Gerd Binnig and Heinrich Rohrer at IBM Zurich, controlling the size of features on material surfaces at the nanoscale was a still-distant goal. The ability to peer "inside" materials was solely done by x-ray crystallography. However, this technique is limited to determining the diffraction of highly crystalline materials; samples that are amorphous or only very barely crystalline find themselves at a significant disadvantage in these techniques. Another method of analysis, neutron diffraction, has been available since 1945, when it was invented at Oak Ridge National Laboratory in Oak Ridge, Tennessee.[29] This technique is extremely expensive, however, and not well-suited to routine analysis. As an added difficulty, both diffraction techniques are better suited to analyzing the larger body of a sample rather than determining what it looks like on its surface. Another technique, x-ray photoelectron spectroscopy, determines the atomic composition of material's surface features. This technique is better for telling you what elements are on top of a sample and poorer at showing you what that surface looks like. Think of it like being able to smell that you have butter on your toast but not being able to see how much butter is on any given point of the toast. It is no simple matter to simultaneously determine what is on the surface of a sample through some signal, and to relay where on a sample that signal comes from.

The Scanning Tunneling Microscope (STM) finally allowed for scientists to translate electronic signals from the surface of a material that is probed by an atomically sharp needle. For STM analysis, a sample is placed inside a high vacuum chamber (a place that can reach almost one-millionth of atmospheric pressure) and, like a record needle detecting the surface grooves of a record, the needle moves across the sample and detects its "grooves." The special thing about STM, however, is that it detects the surface on the atomic level and has been used to show structures and patterns. Modern machines are so sensitive, in fact, that molecules on metal surfaces have been imaged in near real-time, showing the rearrangement of bonds in a molecule undergoing a chemical reaction. While machines in 2017 are not able to act as atomic 3-D printers, a day may come where many needle heads at once act in concert to pick up and place atoms where they need to be, fully realizing Feynman's ultimate atomic machinery.

Computer models have evolved since the 1960s to calculate the electronic structure and shapes of molecules. The models have been modified over time to give better predictions about the behavior of molecules within circuits. Predictions by theoreticians and the results from experimentalists are lining up more and more closely, and this has allowed the field to adapt quickly. The interdisciplinary centers of material science research around the world, like the AMBER (Advanced Materials and BioEngineering Research) Center at Trinity College, Dublin, or the CA2DM (Center for Advanced 2D Materials) through the National University of Singapore, are able to make significant advances in creating and testing new materials. The pace of innovation in these research centers is dizzying.

In 1965, Gordon Moore noticed a trend within the electronics industry that was directly applicable to his fledgling computing company. This trend showed that transistor density on a chip could possibly double every year, which was what led to the immense transition from expensive computing buildings for calculations to the home computers we use every day. The company grew to become the worldwide technology giant Intel. Moore predicted a decade later that the number of transistors on a computer chip would be slower than his first prediction and double roughly every two years. "Moore's Law" has kept pace with innovation through the mid-2010s due to coordination among leading corporations.[30] There is significant discussion, though, about the amount silicon devices can continue to shrink before becoming unusable. Classical models of conduction begin to disappear at the level of nanometer-wide devices, and quantum physics begins to take over. Noise starts to overload the electric signal, and heat buildup kills a device's lifetime. This causes significant added problems that must be solved, perhaps by opening up new avenues of research within chemistry and material science. Meanwhile, Intel and other technology companies announced in 2017 that they would pursue 7 nm transistor technology to incorporate within devices. How much smaller these transistors can get is a matter of intense speculation by researchers within the companies' R&D departments. Silicon or carbon circuitry are still equally affected by this problem. After all, atoms can only get so small before interference or random noise dominates any signal passing through the circuit.

Still, the typical production techniques of photolithography and etching are prime examples of top-down manufacturing, where a large starting material is hewn away into fine components, generating a tremendous amount of waste in the process. Chiseling rock down to a statue or lathing a piece of metal to make tools may not seem like a big deal, but think of all the sawdust generated from making a chair or cabinet in a wood shop. Metal shops are similarly strewn with fine metal shards and rusty dust. On top of this, there are limitations to the shapes that may be built by removing material from a stock block rather than building from the ground-up. Namely, you need to start with impeccably pure material. This is, of course, extremely expensive. Then, you're just removing and throwing away some significant fraction of the stock material that you paid lots of money for, just to get some smaller subsection of that stock. Then, refining the details of your machined parts need to be done with extreme care or else—oops!—you made a mistake that requires the piece be thrown away, and then you need to start again. Additive manufacturing bypasses much of the wasteful aspect of traditional machining, and atomic resolution of the printed product will (someday) give complete control over the thermal, electronic, and optical properties of a designed piece.

While Moore's law has been illustrative in guiding the technology development for microprocessing chips, the other components of modern electronic circuitry have also kept pace in their miniaturization. Wires, diodes, capacitors, and other components have all found ways to miniaturize, allowing for major computing facilities, like that shown in the movie *Hidden Figures*, to give way to the personal computer. The personal computer gave way to the laptop computer. Likewise, the laptop computer gave way to smartphones and tablets. It is the hope of material scientists, chemists, and physicists to pave the way for new, even smaller, devices. It is their hope to move beyond the physical limitations that Moore's Law models, at least for a little while.

Carbon-based devices may have had a rockier start than Professor Inokuchi would have preferred, but they have proven to be a rich area of research since 1974. The Westinghouse Electric Corporation was a major driving force behind the US Air Force's early interest in molecular electronics, but they were never able to deliver on their dream of providing molecular-scale devices. Their goal was to provide these circuits and devices for aeronautic applications, cutting down on the weight of aircraft and increasing on-board computing power. Early focus on lofty language in 1957 and big promises in 1958 won Westinghouse significant contracts in 1959, but they came up short on delivering on those promises. When the partnership between Westinghouse and the Air Force fell through in 1963, a decade of silence resulted for molecular electronics.[31]

In 1974, an Israeli graduate student named Arieh Aviram came to New York University from IBM's Thomas J. Watson Research Center and set out to do what Westinghouse had failed to achieve. Aviram teamed up with his advisor, Mark Ratner, and together they devised the first carbon-based single-molecule diode. (Diodes are used to control the direction of current flow within a circuit.) In essence, they took inspiration from traditional semiconductor design in their approach and applied it to chemical principles to calculate how well this molecular diode would operate.

Their calculations went largely unheeded for over a decade and a half, until instruments were invented that allowed experimental proof of their work—and single-molecule devices became one step closer to reality. In the interim, physical chemists and physicists developed charge transport equations to analyze and predict the properties of nano-circuitry. Graphite-based compounds, fullerenes, and nanotubes were coopted for use within these molecular electronic devices. Suddenly, somewhat quantitative predictions were cropping up in the literature, and molecular electronics blossomed as a subfield of nanotechnology.

Carbon science began to interest a wider group of researchers as the 1970s and 1980s wore on. Research groups expanded to fill gaps in our collective knowledge about the element, and competition bred fierce battles for funding to claim the title of "first" on whatever research idea they could come up with. Tremendous amounts of energy and money were spent to discover and characterize an ever-growing population of molecules based on carbon backbones. In the 1970s, astrochemists (that's right, chemists who deal with molecules in space) detected noticeable amounts of carbon-containing molecules in space. Red giant stars, in particular, contained large amounts of these more complex molecules within their cool, diffuse atmospheres. By looking at infrared emission signatures, carbon clusters were detected in the clouds of gas surrounding red giants, and by reproducing these molecules on Earth in their gas phase, scientists could study how complicated carbon molecules formed in space. These complex carbon-based molecules were of great interest to biologists in particular, because they might contain clues about the origin of life on Earth. As an extension of that, they could provide clues to the possible existence of life beyond Earth. One exciting proposition for extraterrestrial life involves a well-spring of life-bearing planets populated with DNA and protein-using aliens. There is no evidence yet that these organisms exist, but the idea of a universal biology tantalizes the field.

Some of the carbon clusters found around the red giants ended up being fused extensions of benzene, which is a hexagon-shaped molecule of carbon. When you line up benzene's hexagons next to each other so that they share one side, that's the molecule naphthalene (pronounced NAF-thu-leen). If you put another hexagon onto this shape, so that you have three hexagons in a line sharing the two inner sides, this makes anthracene. One could extend the number of hexagons along this line, sharing along the same direction and creating different molecules. Or one could start a second row. Atop the two rings of naphthalene another naphthalene could nest itself comfortably. This molecule is known as pyrene. These molecules and many more like them have been found as components of gas clouds in outer space, although they are also found much closer to home. In fact, they are quite abundant here on Earth. The molecules, because they are made from many interlocking rings of the aromatic molecule benzene, are a part of a class of molecules called polycyclic aromatic hydrocarbons (or PAHs). Chemists that work in the petroleum industry are very aware of PAHs, as they are a major component of coal tar.

Another class of molecules in the interstellar medium contains long linear chains of carbon connected to one another. Acetylene, H−C≡C−H, is the smallest example of this class of molecules. It is two carbons bonded to each other, with each carbon having a single hydrogen bonded 180° from the central C≡C triple bond. If you remove the hydrogen atoms and instead replace that with another unit of −C≡C−, you end up with H−C≡C−C≡C−C≡C−H. This molecule contains many acetylene-like units within the straight-chain structure and is an example of a polyacetylene.

Harold Kroto, from the University of Sussex in the United Kingdom, was interested in studying interstellar chains of molecules ending in −C≡C−C≡N for their interesting infrared signatures. Kroto traveled to Rice University in 1985 to begin a collaboration with Robert Curl and Richard Smalley. These researchers, along with their teams of graduate students, began firing high-powered lasers at a graphite surface with an atmospheric pressure one-millionth that at sea level. Groups of carbon, cooled within this vacuum, were ionized by a high voltage and then analyzed based on their respective masses. Kroto and the others expected to find small groups of carbon clusters. What they found instead was far more interesting. Kroto's analysis was presented in graphs showing the amounts of the different masses against the size of each carbon cluster, enabling the researchers to see how many of each cluster size was made by each laser blast. From this series of experiments, they found a pattern of clusters that they didn't expect. Sixty atoms of carbon had a conspicuously high abundance on the graph, and this puzzled the group. If the graph were a hand, you could say it had many fingers, jutting up from the zero line, but the C_{60} finger was bigger than the rest by a very wide margin.

There were some other oddities in the graph that required explaining as well. The low-mass molecules (clusters of up to around thirty-five carbon atoms in mass) were made up of odd numbers of carbon atoms and contained other elements, H and N, as Kroto had originally predicted. But the high-mass molecules (in this experiment, molecules above forty carbon atoms in mass) were only made up of even numbers of carbon atoms. What shapes could these strange molecules take? For that matter, would any shape necessarily be consistent? For all the researchers knew, these lumps were forming in the machine out of serendipity as much as careful parameter selection on their instruments.

What of that peak for sixty carbon atoms? Nobody could have predicted much from a bump on a graph charting masses versus their relative amounts. It was stable, as proven by its high abundance under a wide variety of conditions in the testing equipment. It was also unreactive with the other elements present,[32] which led to the idea that it did not have any reactive external electrons. Working hard, they tuned parameters within the machine to selectively produce the C_{60} molecule and set about trying to deduce its structure. Boron hydride cage molecules had been known since at least the 1960s, but the hydrogen atoms lay outside the cage structure. This C_{60} molecule, having no hydrogen atoms in its structure by definition, would not be directly analogous to these boron-based cages.

It turns out that in the 1960s, one inventor and chemist had found a passion for writing humorous technical articles for the magazine *New Scientist*. This chemist, David Jones (aka Daedalus, his pen name), published an article in 1966 that treated some of the properties of a hypothetical cage of carbon as a humorous take on a technical publication. He predicted that fullerene molecules, while hollow, would also be empty inside. This would mean that the balls formed would be extremely light. Many of

Jones's predictions were entirely wrong, but accuracy wasn't the point of the article. One fantastical prediction in this article said that the largest-possible hollow carbon shell would have a molecular formula of $C_{200,000}$. One mathematical concept in the write-up was crucial to the Rice University group, although Jones was not the originator of the idea. Leonhard Euler, the eighteenth century Swiss mathematician, developed a theorem that stated that pentagons could be added into a surface of hexagons to close the surface into a polyhedron—a 3-D ball.

Smalley did not initially consider that pentagons could be incorporated into thc structure; his initial attempts at designing the C_{60} structure included only hexagons. He could not build a suitable structure on his computer, though, so he went back to basics and began cutting out hexagons of paper. Taping the pieces together, he tried to bend the shape in a way that made sense, but this approach ultimately failed.

To understand why a material like graphene lies flat with a perfectly regular geometric lattice of hexagons while other arrangements of atoms buckle around themselves into three dimensions, we must consider a bit about ways in which two-dimensional shapes can interact with one another. Perfectly symmetric hexagons, like the structure of benzene and graphene, have high symmetry in their shapes. All sides are the same length, and all internal angles connecting the sides together are the same as well. All carbon atoms in the structure are identical. The internal angles of a hexagon are 120° and add up to 720°, which is important for the hexagons. Three hexagons that share a single vertex (atom) will all lie in the same plane because their total angles around that vertex add up to 360°. Two other shapes, the equilateral triangle and the square, also have internal angles which add up to 360° around a vertex. Six triangles, with angles of 60° each, will tessellate into a flat surface. Four squares will also tessellate with their 90° angles. If you have ever tiled a floor, or watched someone do it, then you understand that only shape combinations that add up to a total of 360° will lie flat.[33] If the sum of angles was over 360°, you have a weird bump in the floor that would be uncomfortable to step on. If the sum of angles were less than 360°, then you would either have a divot or would have to fill in the gap with extra grout.

A YouTube video by the group Numberphile, "Perfect Shapes in Higher Dimensions," illustrates this concept very well.[34] The video shows animations where the regular polygons (equilateral triangle, square, hexagon) fit three polygons of the same type together around a shared vertex. Essentially, one object shares a side with the other two objects, but all three objects must share at least one point. Three triangles that come together at a vertex leave a gap, since their angles add up to 180° and not 360°. The triangles are then able to fold around on one another, creating the perfect shape called a tetrahedron. When three squares fit around a vertex, they fold around on one another to form one half of a cube. The internal angles of a regular pentagon are 108°, so three pentagons around a vertex have an internal angle of 324°. This allows them to pucker into a bowl shape, forming one quarter of a dodecahedron. However, three hexagons fit 360° perfectly, and no buckling is possible. This shape will always be flat. Polygons with more sides than a hexagon cannot have three shapes sharing one vertex because their internal angles add up to greater than 360°.

A pentagon sharing a common vertex with two hexagons on two sides has two options open to it. The pentagon could either stretch and deform itself to have one angle at 120°, and all the other angles would be affected by that, but this would cause the pentagon to lose symmetry and the carbon atoms would become identifiable by differences in the shape's symmetry. The other option is that the pentagon remains a regular polygon, while a gap exists between the pentagon and one of the hexagons. Again, this is not so attractive, since the shape loses symmetry. Fortunately, we live in a three-dimensional universe, and this gives us the option to buckle and form more complicated shapes. Take that, Flatland! This rudimentary bowl-like structure forms the basis of the fullerene curvature. Five hexagons each share a side with the five sides of the pentagon, and each pair of hexagons that share a side also share the same pair of pentagons. The pentagons do not touch one another, they share neither a side nor a vertex.

When Smalley found that he had to include pentagons for the cage to take shape, the problem became much more tractable. The shape formed to close on itself, sixty vertices for sixty atoms, with no dangling bonds, and every carbon indistinguishable from any others. This was the first evidence that Smalley, Kroto, and Curl had found something new, but the fight for acceptance was only just beginning. Working late one night, the group built the very first fullerene model from gummy bears and toothpicks. One can have the best tools in the world, but if those tools still fail then it might be necessary to go back to basics—gummy bears and toothpicks. Only five years after Iijima's paper, and ten years after the discovery of buckminsterfullerene, Curl, Kroto, and Smalley were awarded the Nobel Prize in Chemistry in 1996.

While some people who worked with graphite believed that different shapes of the graphite flakes might be made by rolling or folding the sheets, no experimental proof existed of a sheet existing on its own. As a matter of fact, the prevailing wisdom in science was that no sheet *could* possibly exist on its own. People had done the calculations and pointed out that the vibrations of the atoms within two-dimensional sheets would supposedly shake themselves apart if someone were to try to create graphene (or any other sheet-like material). That was evidence enough for some people through as late as the 1990s.

> Although we have learned much about the phases of carbon [since 1960], much ignorance remains about the phases of carbon, with many new directions awaiting exploration for this fundamental and universally common form of matter.
>
> —Mildred Dresselhaus,
> "Future Directions in Carbon Science,"
> *Annual Review of Materials Science*, 1997

Chapter 3

THE DISCOVERY OF GRAPHENE

The excitement behind the discovery of graphene was not only due to its properties or to its structure. Scientists already knew that graphene would be a conductor; the conductivity of graphite had been measured as early as 1939. What made graphene so exciting initially was how much more conductive it was than had been predicted. Graphene is roughly ten times as conductive as pure iron, and about one and a half times as conductive as silver, the next most conductive metal. Graphene's structure had been known since the 1920s (as mentioned in the first chapter), but the interest came from the derivatives and allotropes that carbon could make, the fullerenes and carbon nanotubes, relatives of graphite. These related materials were discovered in the 1980s and 1990s and confirmed the hypothesis that the graphite edges were rather unstable and reactive. Some critics may have even used the fullerene and nanotube discoveries as a portent that graphene did not or could not exist. It is easy to sit back and think, "If it were possible, we would have found it by now."

Rather, the excitement stemmed from how stupefyingly easy the isolation of graphene from graphite turned out to be. This lead to a collective "Doh!" moment across the scientific community and then a veritable arms race to discover the properties and potentials of the new material. Suddenly graphene became the hottest topic in materials science. The shift happened almost overnight, at least on academic research timescales. While Andre Geim and Konstantin Novoselov were the first to receive widespread credit for discovering interesting electronic properties about single-layer graphene, other groups had been actively working on isolating and measuring "single-layer graphite" (dubbed graphene only more recently, discussed later in this chapter) for some time. Money began to flow, as billions of dollars in research grants were awarded annually all across the globe. While governments, companies, and universities continue to focus their energies on racing products to market with this new wonder material, there is still no guarantee that graphene will succeed. However, the ease with which it had been isolated from graphite ore ignited excitement that might trace its roots in the intercalation compounds prepared by Mildred Dresselhaus in the 1970s. Anyone with some adhesive tape and the right microscope could now find and measure this elusive poltergeist.

In science research, there are different scales of "easy." In chemistry, synthesizing a chemical that you've made a hundred (or a thousand times) before is considered easy, even if it might take you a week or more of grueling labor to do so. If you know all the steps, it's practically rote. On the other hand, if it takes only a day or two of concentrated focus to make a new chemical with well-understood reaction conditions, this could also be considered easy. A few reaction types are especially well known and are named after their discoverers because they have unusually broad applicability or high efficiency. The Haber-Bosch process, Diels-Alder Cyclization, and the Suzuki Coupling are all shining examples of important named reactions used in organic synthetic chemistry. Performing one of these reactions is, on the grand scale of things, easy if you know what you're doing, even if you haven't done the specific reaction before.

This type of thinking is not unique to chemistry. For example, a medical doctor knows to keep the skin around a wound dry to prevent further damage.[1] Of course, someone untrained in the actual process by which this is accomplished might find it difficult, but it is a testament to the education and ability of medical professionals that they treat these conditions without stressing out. Other professions and skilled trades, like electricians or plumbers, each have their own examples of processes that are easy for the seasoned veterans and nearly impossible for outsiders.

The isolation of graphene was so easy that it borders on the absurd. It has since become famously known as the "Scotch-tape method" after the common brand of clear adhesive tape that was first used. The isolation process was actually discovered by accident, despite the fact that a small group of dedicated specialists had already devoted significant attention to the challenge, which had eluded other researchers in the field of carbon-based electronics for decades. In fact, graphene was discovered during some scientific playtime in Geim's lab.

Everyone has their own idea of fun. Some people are content curling up with a good book, perhaps in front of a roaring fire on a cold winter's day, relaxing in peace. Other people enjoy hanging from cliff faces, defying gravity. Other people fish or play competitive video games. For some, though, their profession is their play. Many scientists are like this, and stories abound throughout history of "natural philosophers" dedicating their leisure time (and their or someone else's fortune) to answer profound questions about humanity, the world, and the universe. Antonie van Leeuwenhoek, Tycho Brahe, and Isaac Newton are all examples of these aristocratic scientific minds. (They were wealthy people who also happened to be curious about the world around them.) Passionate and inquisitive minds ask creative questions, with answers that may have little to no immediate economic value, but rather for the sake of the questions themselves. Many of these questions, these aching curiosities, not only keep those asking them awake at night but ultimately serve to transform our daily lives in ways that we cannot immediately foresee.

It is unfortunate then, that the most widely accepted truism in academic scientific culture today is a macabre phrase "publish or perish." While it perhaps began as a tongue-in-cheek allusion to the fact that discoveries are meaningless until they are published and accessible to the wider scientific community, the term has almost taken on its literal meaning. There is immense pressure to out-produce your colleagues (read: competition) by generating meaningful and publishable results that will ultimately bring prestige to one's lab and institution. Funding opportunities, tenure reviews, and academics' self-worth are all tied into the nebulous metrics that determine the value of any given researcher's creative output. The most denigrating of academic advisors verbally threaten the careers of their students if results do not come out to the professors' satisfaction.

Performing experiments "just for the fun of it" are an immensely uncommon privilege, therefore. Scientific playtime is a luxury that few labs can afford. Most are strapped for resources; time and money are always in tight supply. But some passionate researchers can squeeze these supplies into a few more Hail Mary–type experiments. These tests mostly end up being only fun and creative exercises, but occasionally they turn up something unusual or unexpected. Despite being "extra" projects, the experiments are nonetheless controlled and recorded carefully.

The discovery of graphene was just such a project. In 2010, Professor Andre Geim described the discovery of graphene in his Nobel Prize lecture, "Random Walk to Graphene"[2]—the title is a nod to the mathematical idea that things that start at the same initial conditions diverge in their individual paths taken over time because of unpredictable outside influence. The name is also a direct allusion to the fact that this discovery was a product of the famed "Friday Night Experiments," where creative, undirected questions unrelated to the normal research workload were investigated. These questions could form out of nowhere, as random bizarre flashes of inspiration, and the experiments were not necessarily limited to only Friday nights.

The name originated in Geim's first hare-brained idea, which occurred to him on a Friday night while he was working for Radboud University Nijmegen, in the Netherlands. In an NPR *All Things Considered* article, "Ig Nobel to Nobel: Creative (and Fun) Science Wins," Dr. Allen McDonald says about Geim, "He's just exceptionally creative. He's always looking for something new, and wanting to be creative is not enough. He just has tremendous intuition."[3]

This inherent creativity caused Geim to try something daring early in his career. On a late Friday night, he decided that he would pour water into a high-field electromagnet while it was operating. This magnet, a 20 Tesla monstrosity, had a strength of about 400,000x the magnetic field of the Earth.[4] The machine was one of the most powerful electromagnets in the world at that time. While the actual cost of the 20 T magnet was not easy to find, the High Field Magnet Laboratory at Radboud University (where Geim tested his unusual hypothesis) purchased two new magnets in 2014 for a total cost of €2.5 million (just shy of $3.5 million in March 2014 currency exchange).[5] These two new magnets clocked in at an incredible 37.5 T, and remained the most powerful magnets at the facility through 2017. So imagine Geim walking in to the building one night and deciding that he would pour water into this tremendously expensive machine while it was running at maximum power. Fortunately, the cylindrical bore of the magnet passed all the way through the magnet's body, so the water should not have caused an issue and passed right through. Of course, speculation and reality can be two entirely different beasts. When Geim poured the water into the magnet, it ended up becoming trapped in the bore of the magnet, unexpectedly suspended against gravity due to the diamagnetic repulsion characteristic of water.

Diamagnetism is tough to describe. It is a phenomenon in physics that describes how an object placed near a magnet will weakly repel the magnetic field. There are poor analogies for this effect, largely because most of our analogies about magnetism describe the attractive ferromagnetic properties that we observe on our large, everyday scale. It could be said, at the risk of oversimplification, that approaching a diamagnetic material with a typical magnet would repel the material rather than attract it, as would usually happen with common magnetic materials. Wile E. Coyote's ingenious diamagnetic contraptions would push him along the hunt for Road Runner rather than drag him.

Geim's experiment became famously known as the "Levitating Frog Experiment." Geim and a colleague, Michael Berry, were experimenting with the effects of especially powerful magnets on diamagnetic materials as a direct result of the levitating water within the electromagnet. Of course, being creative and hopelessly curious led them to ask a question: If water diamagnetically levitated, and living things were mostly water, could living things be levitated in a sufficiently strong field? As it turns out, the answer is "yes." Geim and Berry were able to successfully levitate a hazelnut, a fish, a strawberry, and, of course, a frog. The videos of the inanimate objects floating and spinning in the magnet are interesting enough, but the video of the frog, with its arms shooting out every which way to find purchase on a surface as it spun around haphazardly, is truly entertaining (the frog was fine when it was released from the magnet). They were able to subsequently publish a paper titled "Of Flying Frogs and Levitrons."[6] Part of their motivation for this work came from a desire to communicate or demonstrate curiosities of science to a lay audience, but this work also won the pair the Ig Nobel Prize for Physics in 2000. Geim went on to become the first person to win both the Ig Nobel and the Nobel Prize when he was awarded the Nobel Prize in Physics in 2010.

In order to understand why winning both prizes is such a big deal, one really must understand what an "Ig Nobel Prize" is and how it may be won. The Ig Nobel Prize has been awarded each year since 1991 in a ceremony that is intended to parody and mock the pomp and circumstance underlying the traditional Nobel Prize award ceremony. Created by Marc Abrahams, a co-founder of the *Annals of Improbable Research*, the ceremony highlights research or activities that are on their surface funny but are also secretly genius. Their website highlights this fact many times over: ". . . honoring achievements that make people laugh, then think."[7] While the awardees for the Nobel Prize are always within certain overarching categories as outlined by Alfred Nobel in 1895 (Physics, Chemistry, Medicine/Physiology, Literature, and Peace), Ig Nobel Prizes are not subject to such limitations; the physical and social sciences are each featured strongly every year. Doctor Peter Barss was awarded the Ig Nobel for research on coconut-related head injuries on tropical islands, chemists have been recipients for research related to the brain chemistry of love and OCD, and the 2016 Ig Nobel Prize in Economics recognized research on "The Brand Personality of Rocks" (whatever that happens to mean). A recent Ig Nobel in Physics revealed that white horses are the most horsefly-proof color of horse. If you laughed at the absurdity of trying to think of someone who would ask such a silly question and why it could possibly be important, then you have found the exact criterion for such a research project to be an Ig Nobel selectee.

Geim was an adventurous, young, independent investigator in the early 2000s; it was fortunate that he was already given extremely wide latitude for intellectual freedom, within an environment that permitted fun experiments. "What if" questions could be floated (sometimes literally!), and if supplies existed that could test the hypothesis, then he would conduct an experiment. When he landed an assistant professorship position in the UK, he carried this model with him. Lab members under Geim at Manchester had some flexibility in their direction of inquiry, and a spirit of cooperation between group members helped overlap skill sets.

Andre Geim and Konstantin Novoselov did not intentionally set out to create graphene, at least not in the sense that they had their eyes set on winning the Nobel Prize. The levitation experiment had taught Geim that poking your nose into research outside of your area of expertise can be an interesting and exciting adventure. It's a mental exercise. Most of the time, these side tracks either end up answering a question someone has already answered or else the trail dead-ends, but either way you grow as a person and as a scientist. While Novoselov and Geim were mulling over a problem that they were having with a recent Friday Night Experiment idea, an unexpected lightbulb went off in Novoselov's mind. A new graduate student in the Geim group had just polished away a block of graphite, looking to isolate as thin a piece as could be managed and, hopefully, eventually turn it into a *transistor*.[8] The group wanted to make transistors out of the thinnest piece of graphite that they could make. This sliver was fairly small, but Geim was unconvinced it was as small as could be physically produced. Work in decades prior had suggested that very thin graphite should produce some extremely interesting physics and this was their shot at contributing something useful.

Unfortunately, the graduate student had used the whole slab of graphite and polished away the whole piece down to one small speck. Geim later recalled the story during his Nobel lecture, describing how this student had "polished a mountain to get one grain of sand."[9] He went on to describe how he had accidentally given the student a block of high-density graphite, a type of graphite that has many different crystal phases all packed together. This kind of material is not as appropriate to use as the highly oriented pyrolytic graphite discovered in the seventies. Pyrolytic graphite has many larger crystals within the structures, which would have made polishing it easier.

The golden moment came soon after, when Geim was speaking with a colleague about the trials and tribulations of getting very thin, very high-quality flakes for the tests they wanted to run. The colleague, Oleg Shklyarevskii, was familiar with ways in which graphite was prepared before being analyzed by a Scanning Tunneling Microscope (STM).[10] Shklyarevskii showed Geim how STM microscopists cleaned graphite samples for analysis by taking a piece of sticky office tape, pressing it onto the surface, and then carefully ripping it away. The tape took away finger oils, dirt, and other grime that otherwise contaminated the surface and would clutter the resulting microscope image. It is like a bikini wax, but for a rock. The tape was just thrown away in the wastebasket. Nobody had thought to actually look at the residue under the microscope, presumably because they thought the image would be cluttered and indiscernible. To the surprise of those eyes peering through the microscope, the flakes of graphite on the Scotch tape were in fact thinner than the polished mountain. With a hint of humor, Geim recalls in his Nobel speech, "Only then did I realize how silly it was of me to suggest the polishing machine. Polishing was dead, long live Scotch tape!"[11]

If you are saying to yourself right now, "Now wait a minute. That seems overly simplistic. I feel like I could do something similar right here and now with a pencil and tape on my desk," you would be right; you could easily perform an analogous experiment yourself. Go and get a piece of paper, a pencil, and some tape. Seriously. Go ahead. We will wait.

Now that you have them, take your pencil and rub about a square-centimeter (or a square-half-inch for the Americans) onto the paper. It doesn't have to be super dark; you can hold the pencil with your typical writing pressure. This is just so you get a good enough surface area that the tape can access. Note that pencil points don't work especially well for this demonstration. Once you have your square drawn as you see in figure 3-1 below, take about a ten- or fifteen-centimeter (about six inches) strip of tape in your hands and hold it between your thumb and forefinger so that the tape is straight and taut. This is mostly so that the tape doesn't stick to itself, to you, or to other stray things.[12] Now, carefully take one end of the tape, the choice of end is up to you, and let about three centimeters (roughly an inch) touch the paper. Press just hard enough so that the tape contacts the whole graphite square but nothing else. Carefully pull up on the tape and remove it slowly so that you're not also ripping the paper. Can you see how some of the graphite transferred from the paper onto the tape? The graphite that transferred onto the tape is somewhat dark, many of the flakes that ended up on the tape are not graphene but are rather small lumps of powdered graphite. We will cleave some of those lumps in just a second. This process of pulling graphite flakes from a surface (whether it be a hunk of graphite or a piece of paper) is called *stripping*, as you are pulling something (in this case surface graphite flakes) off of something else (here, paper).

To simulate the method that produces thinner and thinner flakes, you'll need to do repeated strippings along the length of the tape, folding it and separating the ends along a progressive line on the tape, so that the graphite is divided again and again by the tape's adhesive. Figure 3-1 shows a progressive stripping producing lighter and lighter patterns. There are four total strippings in this picture on the left, but you can only see three. The lightest one would contain a low density of graphite flakes, and likely, would also contain some graphene sheets. Not bad for a few minutes of work! And compared to the pictures of the Novoselov and Geim *Science* article, this is actually a fair proxy.

To view the experimental results and find out how many single-layer graphene flakes you created, you would take a silicon wafer coated in 300 nanometers of silicon dioxide (aka glass) and press this tape with the lightest bit of graphite down onto the wafer's surface. The graphite flakes will stick to the wafer when the tape is pulled off, and a special type of microscope called an interference micrograph (which takes pictures to show topographical differences in small windows) would show single-layer graphene sheets as blue shapes on an otherwise pink backdrop. Few-layer graphene stacks, up to about five sheets, fade to a gray-green color, and bulk graphite sheets of over ten layers or so become yellow. This easy-to-identify scheme was perhaps the most important discovery in early graphene research, as we will discuss below.

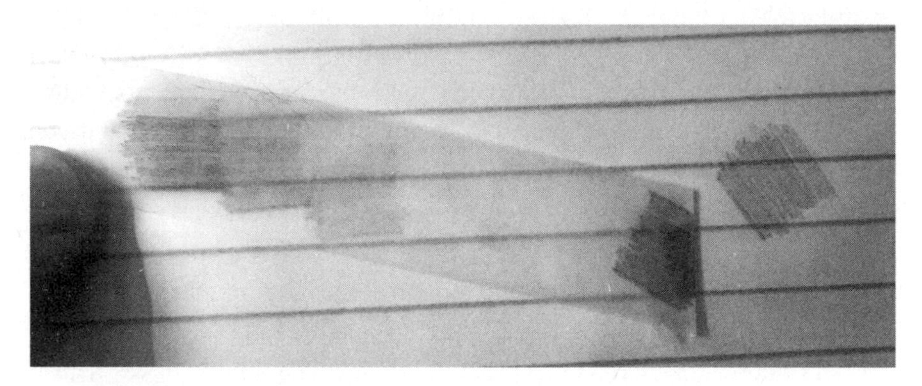

Figure 3-1: A pencil rubbing on paper (right) will deposit graphite material onto a piece of tape. Successively stripping the tape deposit produces successively lighter transfers, showing ever thinner flake thicknesses. (Image by Joseph Meany.)

For the Geim group, the interesting work was just beginning. Pulling some schmutz off of the surface of a graphite lump is nice, but science needs numbers. It needs repeatable data. A paper needs a beginning, a middle, and an end. More than that, it also needs descriptions of interesting new results. These flakes qualitatively seemed thinner than the polished sample—but how much thinner were they, and what could they actually say about the stripping process? Was graphene going to be anything interesting, as researchers had hoped? How many of the theoretical predictions were right, and how many would be overturned?

Konstantin Novoselov, a graduate student in Geim's research group at that time, decided to take on the challenge of figuring out how to best handle their discovery. While he initially performed experiments on glass with flakes transferred manually by tweezers, Geim ordered some silicon wafers whose surfaces had been coated with a thin layer of silicon dioxide. Geim purchased these wafers because he wanted to run some experiments on the electrical properties of graphene on a wafer, hoping that they might find something interesting. The difficulty remained, however, that they required proof. They needed to find a few specimens of graphene that were only a couple dozen layers thick, and they needed to show, with great confidence, that the process was repeatable and robust.

Graphene that thin is essentially transparent to normal light, which is useful in some applications, but only if you're not trying to use visible light to actually see the flakes. Fortunately, Novoselov thought of an ingenious little trick; with the graphene pressed onto the silicon wafer, he used the interference pattern of light waves contrasted between the flakes and wafer to find different flake thicknesses on the silicon wafer. The interference patterns produced different colors for different flake thicknesses. This allowed him to quickly judge the relative sizes present on his microscope image. It was a fortuitous accident that graphene produces such a vivid color pattern for easy discernment. In his 2007 review of early graphene work, Geim mentions that even choosing the wrong thickness of silicon dioxide coating on the silicon wafer would have had disastrous consequences for their discovery; only a short deviation of five percent would have rendered single-layer graphene invisible to their microscope methods.

"The critical ingredient for success," he wrote, "was the observation that graphene becomes visible in an optical microscope if placed on top of a Si wafer with a carefully chosen thickness of SiO_2 owing to a feeble interference-like contrast with respect to an empty wafer. If not for this simple yet effective way to scan substrates in search of graphene crystallites, they would probably remain undiscovered today."[13]

Novoselov's method of finding the thinnest sheets of graphene, developed in the early days of research, could only be appropriately described with the "needle in a haystack" analogy. Think about it. If you want to find a single crystallite of graphene that has been deposited by tape on a piece of silicon, you're going to be hunting among a lot of larger and more disordered pieces. It took patience and perseverance to be able to locate a flake with the right properties. Those were the flakes whose largest dimension were only as wide as a hair. Finding that needle, though, was still only part of the challenge.

Geim had become especially interested in graphene because of its tantalizing, but at that time theoretical, electronic properties. Physicists had been predicting for sixty years that graphene could be exciting, perhaps being an elastic and flexible superconductor. It was up to someone better at the lab bench than at the blackboard to figure out how to verify or refute those predictions. Therefore, Geim's primary interest was in hooking up electrodes to graphene and finding out how well it conducted electricity in a circuit. But how do you clip wires to something that is an insignificant fraction of the size of a hair? As you can imagine, this is not a trivial task. Novoselov was up to the challenge. He used tweezers, a toothpick, and a conductive silver-based paint to carefully draw contacts onto a thin graphite flake. The word "graphite" here was used with intent. The first circuit they made was with a thin flake that was not a monolayer—just one of their very thin pieces of graphite a few tens of layers thick. This rudimentary circuit would have probably been thrown away and the whole project abandoned if they had not seen even a flicker of gain from the small field effect transistor. Fortunately for all of us, they noticed that this little speck had a small but repeatable bump in conduction when the silicon wafer was turned on. What a relief it must have been to be vindicated in this way, and have a hunch pay off in such a clear way.

It should be emphasized here that the 2010 Nobel Prize was not won for isolating or observing graphene. It was not even awarded for the Scotch tape method, as ingenious as it was. Instead, it was officially awarded by the Nobel Foundation "for groundbreaking experiments regarding the two-dimensional material graphene."[14] This is a salient point; Geim freely acknowledges that other researchers had observed thin films of graphite prior to this project and that a few were also likely able to observe monolayers of graphene:

> In graphene literature and especially in popular articles, a strong emphasis is placed on the Scotch tape technique, and it is hailed for allowing the isolation and identification of ultra-thin graphite films and graphene. For me, this was an important development but still not a Eureka moment. Our goal always was to find some exciting physics rather than just observing ultrathin films in a microscope.[15]

Geim and Novoselov's famous 2004 paper in *Science* was rejected by the journal *Nature* twice, for not moving scientific inquiry forward enough. The editor who rejected the paper was likely aware of the previous work done with thin flakes of graphite, and Geim freely admits that a purely observational paper would have been unimportant by itself. What made the 2004 paper stand out, along with several papers that supported the preliminary results, were the measurements showing that graphite flakes were able to change their electrical properties based on an electric field applied to the silicon wafer, in the same way that a transistor would act as a gate. The hard work was just beginning. Measuring the field effect on this one small wafer in a jerry-rigged circuit was definitely quite the accomplishment, but the data needed to be better if it were to attract the attention of journal editors and the broader field. Nevertheless, as the 2004 paper was published in such a high-impact journal, the stage was set for a race to discovery.

Creating an electric field using the silicon wafer as a transistor gate with the two wires become a source and drain across the graphite flake. This bears some explanation. Within a transistor, there are three sites that must act in concert for the device to work. There is a source, a drain, and a gate. The gate acts as a switch by which electricity can flow from the source to the drain. When the gate is off, no electric field is applied to the system; the transistor is also said to be off. Electricity flows through the source to the drain when the gate is on or open, and no electricity flows when the gate is off. It is this on-or-off, 1-or-0 dichotomy that gives transistors within computer circuits their logic and computing ability. Mathematical operations are carried out by electrons whizzing around in circuits at high speeds, carried by metal wires and within semiconductor chips. This electric field switching current on/off within a transistor, is called the field effect.

Some materials, regular conductive materials like wires, are able to conduct between the source and drain normally when the gate is off. Graphene is a conductor at temperatures that humans are traditionally used to working with, and so it would be relatively simple to induce electrons to flow in a graphene-supported circuit. What was so surprising to Geim and Novoselov, even in their earliest experiments, was that a graphite flake several nanometers thick could exhibit a boost in conduction when the gate voltage was switched from off to on. Not only that, but it didn't especially matter what direction the voltage was applied. If the source and drain wires were switched, *the same field effect was witnessed across the microscopic device*. While it may seem like a ridiculous idea that graphene flakes would somehow not be ambipolar conductors due to its simple chemical makeup, it was proper to confirm that fact. Electrical devices made from graphene would not need any added special handling, which would drive up the barrier to consumer adoption. As this effect was witnessed without regard to the direction of the electricity flow, it is dubbed the ambipolar field effect (ambi- being the prefix for "both," like in ambidextrous; and -polar, referring to the poles of the electromagnetic field).

It took Geim and Novoselov several months to write up their data and submit it to a journal for publication. In October 2004 they published a paper in *Science* titled "Electric Field Effect in Atomically Thin Carbon Films."[16] This paper was followed up in 2005 by two more papers in other scientific journals, "Two-Dimensional Atomic Crystals,"[17] and "Two-Dimensional Gas of Massless Dirac Fermions in Graphene."[18] Together, they provided enough evidence to give them hope that pursuing finer, smaller, and more carefully fabricated devices would be worthwhile. They turned their attention to collaborating with other researchers, expanding their capabilities to produce higher-quality devices that could be more readily characterized. Complicated experiments that serve no further purpose than to confirm a device's proper manufacturing are expensive, more so if you find devices that were not made properly because then you would have to trash the device and start again. Better manufacturing processes let you get back to performing more new experiments. They were then able to see that graphene monolayers were extremely interesting and that they were onto something big. In perhaps the understatement of the decade, Geim said that, "It is not the observation and isolation of graphene but its electronic properties that took researchers by surprise."[19] He is right, and the electronic properties were the most interesting aspect about graphene prior to researchers probing its mechanical properties.

Konstantin Novoselov realizes full well the importance of his discovery and its huge economic possibilities. Still, he is a champion of science and of the democratization of sharing information for the exhilaration and rush of discovery, even if he may not be making all the discoveries. "I think that's the right scientific style, to share the results openly with other labs," he said in an interview sponsored by the University of Manchester.[20] We know so much about the wondrous properties of graphene largely because researchers can experiment, share, debate, and therefore grow our collective understanding rather than funnel the knowledge to only a select group of engineers.

This echoes a sentiment from an English chemist and inventor from two hundred years before Novoselov, Sir Humphry Davy. In chapter 2 we described how Davy created a lamp from charcoal. As it turns out he was also quite the wordsmith. His opinion on the exchange of information could easily be read as modern-day scientists espousing the benefits behind the modern Open Source Data movements and predicting the Dunning-Kruger effect in the same breath:

> As in Commerce, so in science, no country can become worthily preeminent, except in profiting by the wants, resources, and wealth of its neighbours. . . . Fortunately Science, like that nature to which it belongs, is neither limited by time nor by space. It belongs to the world, and is of no country and no age. The more we know, the more we feel our ignorance, the more we feel how much more remains unknown.[21]

When Geim said that the observation of graphene flakes was not the defining moment of their discovery, it served a purpose other than simple humility. Geim and Novoselov's claim to the graphene throne, as endowed by the Nobel Foundation, did not proceed without its own level of controversy, as research in carbon nanomaterials had already been progressing at a breakneck pace for a decade and a half. Carbon nanotubes had been in the spotlight for nearly a decade, and it was well known by 2004 that opening/unzipping nanotubes would yield a pristine graphene sheet. As can be shown from the scientific literature, characterization of graphene was already a hot area of research when the pair entered the field.

For example, two months after the first Novoselov and Geim paper in 2004, Professor Walter de Heer at the Georgia Institute of Technology published a paper, "Ultrathin Epitaxial Graphite: 2D Electron Gas Properties and a Route toward Graphene-Based Nanoelectronics," which described a different preparation for graphene sheets—de Heer and his coworkers grew graphene from silicon carbide. This type of graphene, called epitaxially grown graphene, is related in part to the graphite synthesis pioneered by Edward G. Acheson in the late 1800s. We will learn more about him in chapter 5 when we talk about the commercialization potential of graphene. Professor de Heer had been working on carbon nanomaterials since 1993, and moving from investigations of fullerenes to nanotubes seemed to be the perfect set-up for an early graphene expert.

De Heer wasn't the only person studying graphene early on. De Heer noted in a 2011 review article that A. J. Van Bommel was the first to pioneer this new generation of SiC grown graphite sheets, and Van Bommel was also able to characterize the sheets on a surface.[22] Van Bommel is quoted as saying in his article that he identified "monocrystalline graphite monolayer films."[23] Since "graphene" as a term was not invented by Hanns-Peter Boehm until 1986, early studies concentrated on some term related to monolayer graphite. Boehm used reduced graphene oxide flakes suspended in water to deposit flakes onto a Transmission Electron Microscope grid in 1962. This paper by Boehm and his coworker Ulrich Hofmann is widely considered to be the very first observational report on graphene. Its title, "The Adsorption Behavior of Very Thin Carbon Films,"[24] is a sober understatement of what was to come.[25]

Part Two

INFILTRATING OUR LIVES

Chapter 4

A MIRACLE MATERIAL WAITING TO BURST FORTH

If graphene is made from carbon and scientists have known how to isolate the material for over a decade, why are there so few graphene products on the market? We are still waiting for our hoverboards, our faster-than-light spaceships, and our glowing *Tron*-like bodysuits. The roadmap from a fundamental research laboratory to store shelf is never a direct path, although the time that passes between discovery and commercial application is shrinking rapidly. Commercialization of electricity and the combustion engine allowed for an unprecedented explosion in innovation across multiple fields. More than ever before, it is easier to move a human around the globe. Nearly instantaneous exchange of information all over the world allows faster collaboration between widespread individuals. Access to information has democratized invention for most of the developed world, and we are steadily seeing an increase in the quality of life for most of the world. Ultimately, graphene will be available for all to use. It will power our houses, it will clean our water. But where is it *now*?

The roadmap for the development of commercial products from graphene might well follow a similar the roadmap as the commercial development of aluminum. Hans Christian Ørsted is credited with discovering aluminum from alum powder in 1825, beating out Humphry Davy (him again?). Davy is recognized as an early contributor to aluminum's ultimate isolation, but his experiments did not produce an appropriately pure sample for full characterization. Ørsted's work was followed up by Friedrich Wöhler in 1827, who crushed vitalism theory (see page 24) a year later with his synthesis of urea from a mineral. The similarities between aluminum metal and graphene share a deeper parallel, though. The mineral alumina (Al_2O_3), like graphite, had been known and used since ancient times. Yet, the true properties of both lay hidden until scientists had unlocked the secrets of nature through sheer force of will.

Aluminum's commercialization did not begin with airplane hulls or foil to protect our leftovers. In fact, the metal was so difficult to produce and work with that it was designated as a precious metal. The metal was so precious, in fact, that the tip of the Washington Monument in Washington, DC, is made from aluminum. You read that right. While statehouses and royal palaces across the globe find themselves adorned in gold, the memorial to America's first president is capped in a lustrous yet colorless metal, inscribed on all four sides with important dates and names. Aluminum utensils found use at the table of Napoleon Bonaparte. For over sixty years it was extremely difficult to produce pure aluminum, and for that reason it was the most expensive pure metal in the world. Aluminum-oxygen bonds are very tough, and this requires a significant amount of energy input to break the bonds and produce pure aluminum.

Aluminum, however, was not going to be so easily coaxed from solution. This energy could not come from fires. Flames are not hot enough to impart the right amount of energy to separate these two atoms from one another. Instead, aluminum will only be purified by high temperatures within electrically driven furnaces. Electricity had been successful in purifying many of the alkali and alkaline earth metals (the first two columns of the periodic table) early in the 1800s, so aluminum purification through electricity would become a natural extension of these investigations.

It was not until a twenty-three-year-old student named Charles Hall took on a challenge set forth by his professor at Oberlin College, Frank Jewett. Batteries and electric turbines were much better in 1886 than they were in 1820, so power generation was much more accessible for these types of high-energy reactions. Aluminum metal has a spare empty orbital, which makes it extremely reactive. Early attempts to create aluminum on a massive scale failed because of the water solutions in which they were carried out; the aluminum would react with the water and become aluminum oxide. To get around this, Hall figured out that he had to adapt methods used to electrolyze other metals (such as magnesium or calcium), and this required some incredible ingenuity. It involved dissolving aluminum oxide in a solvent, but it could be no ordinary solvent. He dissolved the aluminum oxide in molten sodium hexafluoroaluminate—better known as cryolite. He heated the cryolite to about 1000°C within his electric furnace and, when it melted, he dissolved the aluminum oxide within it, also adding some aluminum fluoride to lower the temperature of the melt. The crucible and rods used for adding electricity to this mixture were made from none other than graphite. Electricity delivered to the molten solution would cause the aluminum-oxygen bonds to break, and aluminum globules could then be collected from the reaction.

Paul Héroult, a French chemist also interested in electrolysis, demonstrated essentially the same process as Hall around the same time. Alexander Graham Bell and Elisha Gray may be one of the most famous cases of patents for similar products being filed at the same time, but Hall and Héroult certainly make a top five list of simultaneous inventorships. Hall and Héroult were both awarded patents in Europe and America, and today the Hall-Héroult process continues to produce the world's smelted (non-recycled) aluminum. Héroult went on to invent another important related process, inventing the electric arc furnace for smelting iron. This process uses graphite rods as well, to create the electric discharge and heat metal to its melting point for casting.

After the Hall-Héroult process reduced the cost of producing aluminum, the price of aluminum products proportionally decreased. We encounter it every day: when we open an aluminum can, our Apple computers are encased in it, and the wheels of NASA's Curiosity rover on Mars are made of machined aluminum blocks. What was once a nearly priceless precious commodity, has become commonplace and ubiquitous.

Graphene products will follow the same trajectory. The graphene flakes on silicon wafers are really just the first droplets in the bottom of a beaker when compared to the revolution that will occur once someone solves the riddle of making large-area pristine graphene sheets. We have known about graphite for millennia, and we have finally come to realize its true potential now that we can examine its qualities. Right now, high-priced goods with moderate quality exfoliated graphite samples are hitting the market for sale, and this will generate the revenue necessary to continue research in cutting-edge applications. Once atomic control of production is realized, then the price of graphene will plummet. Even with lower prices, however, the number of products that will use graphene will explode after that point, generating incredible amounts of money.

There is a path through the wild, wild west that is the graphene production industry. Processes are being refined and applied to manufacturing our wondrous carbon friend in bulk, with new methods being discovered regularly. The price continues to drop, so that tinkerers and researchers the world over can experiment and find new applications for its use. Once we have graphene available in affordable mass quantities, how might it be used to change the way we make things?

For the last decade or so, Additive Manufacturing (AM) has been all the rage. You might know AM by its more common name, 3-D printing. Hobbyists use the latter term, researchers and industry tend to use the former. For the purposes of our discussion, they are one and the same. AM describes a process by which real-world, three-dimensional objects are built by adding layer upon layer of material, nearly any material. Many early generation AM devices used only plastic, to make interesting 3-D renditions of various objects, but the technology has grown significantly more capable, with many more materials being used to create not only physical mechanical objects but also functional, complex machines that now rival traditionally manufactured ones in performance and lifetime. AM devices are often called printers and, like printers, they take their instructions from a computer. In this case, the computer has the design of the object to be built defined in great detail using state-of-the-art computer-aided design, or CAD, software. Once a CAD design is produced, the AM equipment reads in data from the CAD file and lays down successive layers of raw material in a layer upon layer fashion, to fabricate a 3-D object.

Additively manufactured structural materials are an obvious place to begin adding graphene flakes. Researchers at MIT, using a custom AM machine, printed various 3-D objects from graphene and tested them to measure their physical properties compared with comparable, more conventionally produced parts. The results were astonishing. Some of the 3-D printed samples had ten times the strength of steel at one-twentieth the mass.[1] They can now print parts and assemblies that may, in some cases, replace custom manufactured steel parts for increased mechanical strength.

AM devices can now also make more complex systems like engines, with moving parts and many, many fewer individual components than the originals, since the components previously had to be manually integrated into the final product. It is interesting to note that this technology is being aggressively pursued in just about every industry, including space exploration. It has been widely reported that rocket manufacturers like Boeing, SpaceX, and Launcher One are using AM to build part of their rockets. There are problems, of course, because not every material needed to make some products are (yet) compatible with AM processes. For example, devices with integrated complex electronic circuitry are not yet able to print at the micro and nanoscales required. Granted, there are other processes in development, and in some cases already being fielded, that "print" complex electronic circuits. The integration of these processes with the structural and mechanical systems produced by mainstream AM devices is not yet perfected and still requires traditional handling and manual (human or robot) processes. Molecular electronics, discussed in chapter 2, will use pre-planned chemical principles to create these complex three-dimensional circuits. Graphene, with its superior heat conduction properties, will help keep these circuits cool within the structure. We are not yet able to make everything, anytime and anywhere, but this is the ultimate dream of many in the AM community. Imagine printing a house whose wiring, plumbing, and heat/AC were just as seamlessly integrated into the structure as different colors are integrated within a color printout.

Graphene will help us take the next steps toward these goals. For example, 3D Graphene Lab, Inc. sells a conductive graphene polymer filament.[2] In other words, they are manufacturing *electrically conductive plastics* that can feed through a conventional 3-D printer as a step toward the integration of structure and electronics. The logical outgrowths of this technology are printable optoelectronics, capacitors, transistors, and other sensors that have been discussed in this book, all enabled or enhanced by graphene.

While small-scale laboratory efforts successfully produce minute quantities of graphene, however, scaling up production to amounts needed for commercial application is a challenge, with long-term storage and transport also hampering efforts. If you do a simple online search for "buy graphene," you will find multiple companies willing to sell you a bottle containing a black powder and calling it graphene. Unless there is strict quality-control testing behind the production methods, though, you can't be sure that what you are expecting is what you are getting. As graphene's superlative qualities come from the carbon-carbon bonds within a monolayer, it is incredibly important to keep in mind that graphite flakes can be hundreds of layers thick and still be a nanomaterial. While these stacks may still be useful for producing some new materials, the truly exciting possibilities for graphene as a revolutionary material will stem from monolayer graphene being incorporated with precision.

When I (author Johnson) was in elementary school, I had a chemistry set. This wasn't one of those wimpy and *safe* chemistry sets for sale today. No, this was the real deal that contained reagent bottles of tannic acid, cobalt chloride, sodium ferrocyanide, and my personal favorite, phenolphthalein solution. Plenty of glass test tubes and beakers came with the set, along with a sample of uranium ore (!!). Remember, this was the 1970s and before we decided that having our children poison themselves was not a good idea. It was kits like this that reinforced my lifelong interest in science and led me to choose chemistry as one of my college undergraduate majors.

Little did I know, as I was using these wonderful (and sometimes carcinogenic or toxic) chemicals and diligently taking notes in my laboratory notebook that I was inadvertently synthesizing a twenty-first-century wonder chemical now known as graphene. I wasn't making it with my chemistry set, though, but with my lowly number two lead pencil as I scratched my observations on the chemical-stained pages. If only it were that easy to make graphene in usable quantities.

Industrial processes to make or isolate specific chemicals can be intimidating. Just consider an oil refinery. To get from the plain, black, molasses-thick crude oil that flows from the ground to the gasoline that goes into your car or the plastic used to make water bottles is a relatively complex process that involves huge machines, high temperatures and pressures, scary sounding chemicals, and a huge risk to the health and safety of refinery employees. The first step is distillation, in which the crude oil mixture is heated in a tall tower (the distillation column, or still) to separate out the heavier carbon molecules, which sink to the bottom, from the lighter ones (like propane), which float to the top. The molecules that condense in the middle of the still are later converted to the fuel for our cars and airplanes. Each layer is separated from the others, sent along separate pipes for different destinations. Nothing is left to waste, and each component molecule has a predetermined fate. Top layers are very light hydrocarbons (methane, ethane, propane, butane) that are converted into liquid and stored. Other light hydrocarbons with reactive structures (ethylene) are diverted to make plastics or more complicated chemical building blocks. The bottom layer, or leftover hydrocarbons in the still are mostly tar or asphalt-like, thick and viscous. The middle layer of sludge is subjected to high pressures and hydrogen gas to make gasoline, diesel, and airplane fuel in a process called conversion. Finally, this impure gasoline is treated with chemicals to remove contaminants like sulfur and nitrogen before it is sent to storage facilities, trucks, and eventually service stations where we refuel our cars. The facilities to do this can require square miles of area and employ hundreds, if not thousands, of people.

By comparison, the method used to isolate graphene from ordinary graphite using pencil lead and tape sounds mundane, and too easy to be real. And, as it turns out, it is. For graphene to make all the revolutionary changes that are predicted (and, in some cases, actually tested), there must be an automated manufacturing process to produce kilograms of graphene per day or tons of the material per year—not just a few grams here and there. As we described back in chapter 1, graphite is basically graphene layered upon itself, waiting for someone to separate it out. This is where it gets tricky, however.

First of all, we should probably rule out mass production of graphene using the method by which it was originally isolated. While it is amusing to imagine a cavernous room filled with people using adhesive tape to separate graphene sheets from piles of pencil lead, it is simply not practical. Perhaps someone can figure out how to automate this particular process, but, even then, it doesn't appear likely to scale well to the mass production needed. In other words, don't invest your retirement savings in adhesive tape futures!

Researchers at Rutgers University are making sheets of graphene out of ordinary graphite flakes and some sulfuric or nitric acid. These acids have scary reputations, thanks to movies and TV, but they are actually quite common and used regularly in chemical processes all over the world. The addition of the acid oxidizes the graphene sheets that make up the graphite, forcing oxygen atoms between the sheets of graphene causes them to split apart, forming graphene oxide sheets suspended in acid and water. Next, the liquid is filtered out, leaving flakes of graphene oxide to clog up the filter. The sum of all the clogs across the filter eventually makes up a paper-like sheet of graphene oxide. This paper-like sheet can then be "removed" from the filter by dissolving the filter away using a solvent that doesn't react with graphene oxide. The last step is to remove the oxygen, which is done by using hydrazine, leaving only a pure graphene coating.[3]

This resulting material is called *reduced* graphene oxide, or RGO for short. In this instance, "reduced" refers to a chemical use of the word, where the *oxidation state* of each graphene carbon has been decreased through the removal of the oxygen by hydrazine. In this case, hydrazine is a *reducing agent*, which is oxidized by its reaction with the graphene oxide.

Many interesting chemical reactions happen when you put energy into molecules. We humans learned this long ago as we built bigger and hotter fires to smelt different metal ores into the metals that underlie our civilization. We tweaked what we burned, the shape of the furnace used, and the amount of oxygen required to make the fire "just right." Heating can also be used to make graphene, using Chemical Vapor Deposition (CVD) as described in chapter 2.

Methane, a carbon-rich gaseous compound with which we humans are very familiar, can be reacted with copper at high temperatures to produce graphene. Simply heat the copper to about 1000°C and expose it to the methane gas.[4] Layers of graphene will be formed on the copper's surface from the plentiful carbon atoms in the methane gas. Here are two big problems with this method: 1) it takes a long time to make even a little graphene and 2) the quality of the graphene produced is not very good.

Dr. David Boyd at Caltech, along with his research collaborators, has found a way to improve on the CVD approach so it will work with lower temperatures and produce a higher quality graphene. They, too, use copper and methane, but they add a bit of nitrogen to improve the layering of the graphene on the copper. In this method, energy still needs to be added, but not nearly as much. The reaction goes forward at a "mere" 420°C. Global industry has considerable experience with CVD, so it should be possible to eventually automate the process on a large scale; the goal is to produce inches, feet, or even yards of high-quality graphene at a time.[5]

Are dangerous chemicals, complex machines, and multistep chemical reactions and processes too complex for your tastes? Then consider this approach, discovered at Kansas State University, where they produced graphene by creating an explosion.[6] Have you ever built a spud gun? Basically, if you take a one to two meter long PVC pipe, create a combustion chamber at one end using a spark plug and a quick-sealing endcap, stuff a potato in the other end and fill the now sealed combustion chamber with a flammable vapor (hair spray is good), then you have a spud gun. Once the potato is in place, the chamber fueled with hair spray and then sealed, you can point the far end of the PVC pipe toward your target and, discharge your battery to cause the spark plug to spark. The resulting small explosion creates a pressure wave that dislodges the potato from the end of the combustion chamber, up the nozzle of the PVC pipe and into the air—often launching it tens of meters into the distance. The physics of what happens in the combustion chamber is very similar to the method that scientists at Kansas State University used to create graphene, in what may become a scalable process that will be a step toward mass production.

Instead of PVC pipe, the scientists used a more robust chamber for their combustion event. They replaced the hair spray with acetylene or ethylene gas mixed with oxygen. They did use a spark plug to create the combustion, just like we did with our spud gun. The fuel, the acetylene or ethylene gas, was turned into graphene and some other carbon detritus.

Interestingly enough, graphene wasn't what the scientists were trying to make. Instead, they were trying to make something called a carbon soot aerosol gel. It is easy to see how this process might produce soot, but useful soot? That's where the idea delves into the university's patented system for creating carbon soot aerosol gels for use in insulation and water purification systems—the *raison d'etre* for the Kansas experiment. These gels were suddenly forgotten when they realized that their soot wasn't what they were looking for, but graphene. And not just a little bit of graphene. They claim that their process is the least expensive so far for potentially mass-producing graphene, and that it doesn't require much input energy.[7] Granted, nothing is ever that simple, but this approach sounds like a good one to pursue in conjunction with other methods.

Then there is the soybean oil CVD method for producing graphene. Yes, soybean oil. As in the same stuff you use at home when you cook.

Do you get the theme here? People all over the world are coming up with creative new methods to produce graphene. Now that they know what they are looking for, they are finding graphene nearly everywhere. A research team in Australia found a way to use everyday soybeans to produce single-layer graphene sheets on top of a nickel substrate—potentially making sheets with large areas all at one time. The process is a variation of the CVD process described previously, but with a significant difference: this one is done in ambient air (no specialized vacuum chambers, etc.) and the required energy is not as great as is required with other CVD processes. The secret is in the nickel foil catalyst used and in carefully controlling the temperature of the process to prevent, as much as possible, the formation of carbon dioxide. Voila. In goes soybean oil—out comes graphene. It is worth mentioning that the team investigated other metal foils, including copper, and they did not promote the formation of graphene. Nickel did.[8]

When all else fails, why not just go home and use our blender to make the wonder material of the twenty-first century? That's essentially what Jonathan Coleman of Trinity College, Dublin, did when he and his team put some graphite in a blender, added an over-the-counter dishwashing liquid, and hit the start button. With only a little more processing required to separate the newly formed graphene sheets. Coleman and his colleagues found that they could produce several hundred grams per hour using a fairly modest set of mixing equipment in a 10,000 liter vat.[9] It isn't yet clear, however, if this method can provide high-quality graphene.

A search of the scientific literature reveals a myriad of techniques that can produce graphene of varying quality. Most have imposing sounding names like sonication, electrochemical synthesis, epitaxy, and sodium ethoxide pyrolysis. What they have in common is complexity, energy, and the fact that they can only achieve the production of small quantities of graphene, which then needs to be separated out from the other reaction products. To date, there is no simple production technique to result in large quantities of high-quality graphene. For the truly remarkable wonders of graphene to be realized, it must be produced in massive amounts—cheaply.

And that is a goal coming closer to fruition thanks to the innovators who pioneered its discovery and fabrication using techniques mentioned above in addition to others not covered here.

Would you like to buy a 10 mm x 10 mm monolayer of graphene flakes on a silicon substrate? $146. How about a 60 mm x 40 mm piece of monolayer graphene on copper? $172 (in 2017 dollars). There are companies specializing in graphene that will sell individual users samples at very reasonable prices.[10] In fact, for $124 and up they will sell you a small bit of graphene on your own custom substrate.

Making graphene, though, is not trivial. The best mass-market graphene comes from chemically exfoliated natural mined graphite, and companies that own interests in graphite mines are already establishing themselves as players in this graphene revolution, leveraging their preferential access to raw materials in order to increase share prices. This echoes the aluminum market development—take an abundant and cheap mineral and refine it into something far more valuable. But without agreement in the market or regulation, how would a buyer determine which so-called graphene product would be best for their needs?

The Center for Advanced 2D Materials (CA2DM) at the National University of Singapore has established seven different tests by which it measures graphitic materials to establish quality and identity. Unfortunately, only a few of these tests are within the reach of a typical company laboratory; the others require expensive equipment that needs to be run and maintained by specially trained technicians. A company creating graphene in the future would probably have to have all of these tests available in-house to minimize lead time. You can't exactly afford to ship a sample to Singapore every time you need to lot test.

The three cheapest tests to perform determine the size of a particular flake, the degree of defects within a given sample, and the elemental makeup of a sample. The size of a flake is determined by an optical microscope, where a graphene/graphite sample on a backing surface is measured by a typical light microscope. A camera and computer are able to measure the rough dimensions of a graphene/graphite particle and report roughly how big the resulting flakes are.

Since graphene's electronic properties are very sensitive to defects in the flakes, the degree of these defects is an important parameter to measure. This can be achieved by a measurement called Raman Spectroscopy, which measures vibrational patterns in the sample. Oxidation of the carbon-carbon bonds in graphene by oxygen open up graphene to environmental degradation (which we will discuss in more detail later on in this chapter), and the introduction of other atoms onto the graphene surface cause various properties to change dramatically. For example, adding even a single hydrogen atom to the graphene structure causes the graphene to become magnetic.

The defect measurements would be supported by elemental analysis, particularly the Carbon-Nitrogen-Hydrogen-Sulfur (CNHS) analysis. Mined graphite would contain residual elements from the formerly living matter which it was created from, and these elements would ultimately detract from the quality of the graphene through one mechanism or another. Unfortunately, CNHS analysis is a destructive technique. Part of the sample must be burned for the components to be analyzed. While this would be useful for batch-to-batch control of relatively cheap industrially exfoliated graphite, it will not be acceptable for samples of graphene produced by other methods.

There are many ways to determine the number of layers in a given graphite flake. One such test, called atomic force microscopy (AFM), uses a hair-thin needle mounted on a small springboard-like lever to measure the atomic forces between the needle and a sample. A laser reflects off the top of the lever, which is able to measure the amount of deflection, up or down, that the needle experiences in its interaction with the surface. The readout gives the thickness measured, and since graphite flakes stack at a constant distance from one another you can do the math to determine the number of layers from that. AFM is able to create an image from many scans, as it adds successive 1D lines together to display a sample's topography. In effect, it creates a height map of a surface.

Scanning electron microscopy and transmission electron microscopy are methods of looking at what a flake of graphene looks like, but on a much finer level than optical microscopy is capable of. These two analyses have a much higher magnification resolution and are therefore able to find rips, tears, and other punctures in a flake, either naturally existing or that may have formed as a part of its isolation or handling. These two analyses combined with AFM would give the most complete 3-D picture of a graphene/graphite sample overall.

The last major analysis performed by CA2DM is x-ray photoelectron spectroscopy (XPS). XPS determines the chemical makeup of a sample nondestructively, and so would give you all of the information that CNHS provides while still being able to recover your sample. In this technique, x-rays are fired at the graphene surface, and some of the x-rays are absorbed by electrons in the sample. The electrons are ejected from the sample with an energy characteristic of the element in the sample, which tells you what elements are present and in what amounts.

Silicon carbide was an easy entrepreneurial target because the initially envisioned uses for it were comparatively low-tech. Simple abrasives do not need to be exceptionally pure to function as advertised. Commercialization did not require a large infrastructure to turn the discovery into a marketable product. Carbon fibers, on the other hand, did not yield a product that could immediately be sold. Instead, fibers required the machinations of a huge corporation to go from "Huh, that's funny" to significant return on investment. Graphene products using the full potential of the material are not going to come from the backyard inventor.

Small companies would be most wise to form relationships with universities or larger companies that are equipped with appropriate instruments. Strategic partnerships (especially by entrepreneurs without professorial jobs) will extend the company's access to fortuitous interactions, as well as to the instruments mentioned in the preceding paragraphs. First-time entrepreneurs can even get development and marketing assistance from a university's technology transfer office. An additional bonus to this relationship comes to the company in the form of an employee pipeline. Top undergraduates, graduate students, and postdocs can easily be tapped for future employees according to the company's needs as it grows. It is a win-win for everyone!

Other than the Scotch tape method and chemical exfoliation, what could our options be for making graphene in large amounts? Is there any way that we might print or grow something into graphene? Mechanical exfoliation (the Scotch tape method) was covered thoroughly in chapter 2. To quickly summarize, adhesive tape may be used to peel hunks of graphite from the surface of a larger graphite hunk, then use successive peelings to isolate a few monolayer sheets. This process has been dramatically improved over the years, and in fact special tapes are now used, which can dissolve in water or other solvents more easily than can office tape. That makes depositing graphene flakes even easier than before. The second method we have mentioned, chemical exfoliation, has a history going back to the late 1800s. As with the mechanical exfoliation process, researchers have added to the field by developing new exfoliation parameters. Generally they are less harsh on the graphite and so minimize damage to the graphene surfaces. Perhaps the method uses recyclable materials, which would be tremendously important for any company that wants to produce literally tons of graphene per year. Some of the improvements improve the yield of pristine monolayer flakes, which is the most important optimization of all. We learned in chapter 2 that highly-oriented pyrolytic graphite (HOPG) allowed Millie Dresselhaus to perform her groundbreaking experiments on the electrical structure of graphite. That HOPG was made by decomposing hydrocarbons (like methane) at high temperature in a furnace through a process called Chemical Vapor Deposition. What similar methods could help us finally produce graphene sheets that will bring us the future?

Graphite production did not always come about through a conscious process. Not all of the great science breakthroughs do. Sometimes, fortunate experimentalists just happen to work on the right areas for new discoveries to happen. "In the fields of observation," said Louis Pasteur in 1854, "chance favors only the prepared mind."[11] Such was the case for Novoselov and Geim in 2004, and such was the case for chemist Edward G. Acheson in 1896.

Acheson only had formal education until he was sixteen, when he left school to earn money for his family working at the Pittsburgh Southern Railroad. He was curious, though, and taught himself after work. He experimented in the evenings, and he eventually built a battery that Thomas Edison bought the rights to. Edison hired Acheson to work at his research lab in Menlo Park, New Jersey, where Acheson worked for Edison from 1880 to 1884, after which he left to become an independent inventor. Sometimes, even working for the best is just not quite as good as working for yourself. Like Charles Hall before him, Acheson acquired a furnace capable of reaching extremely high temperatures. Acheson then began working with creating high-temperature composite materials, mostly in order to synthesize diamonds.

Eventually, while mixing molten clay with carbon under a carbon arc furnace, Acheson discovered granules of a shiny, hard substance among the other products of his reaction. This substance turned out to be silicon carbide, SiC, which has a hardness similar to that of diamond. For this process, in February of 1893 Acheson received the patent for the production of silicon carbide., In reference to the material's hardness, similar to the mineral corundum, he called SiC *carborundum*. He then formed the Carborundum Company and moved to Niagara Falls, New York, to make use of the city's hydroelectric plant. The commercial success of the company brought Acheson into contact with the Cowles Electric Smelting and Aluminum Company in 1900, when he was sued over his use of the electric arc smelting method, which was protected by a patent held by two brothers, Eugene and Alfred Cowles. The lawsuit was settled in favor of the Cowles brothers, although the Carborundum Company continued to produce SiC commercially after paying a royalty to the Cowles' company. A letter in the 1900 *Journal of the American Chemical Society* reports on this case, asserting that the Cowles brothers should be cited as the rightful inventors of silicon carbide saying, "I [Charles Maberry] further asked him [Otto Mühlhaeuser] whether the author was aware that in 1885, the substance to which had recently been assigned the name carborundum, was made in the Cowles furnace, and that specimens of this material could be found in several museums throughout the country."[12] This was in a response to a letter that Mühlhaeuser had published in 1893 detailing the Acheson process and giving all credit to Acheson for using the arc method to create the SiC.[13] The courts decided that, while the use of the arc method was very clearly property of the Cowles brothers, no decision was rendered specifically to cover the carborundum material. The Acheson Process is named to commemorate him for this invention.

The high power offered by the Niagara Falls hydroelectric plant allowed Acheson to continue his efforts to make synthetic diamond. At this he never succeeded, but he did end up producing another unexpected result. In 1895, his experiments created a synthetic graphite, produced when he heated up silicon carbide. He received a patent for this in 1896, and processes that required purer graphite than that which could be mined were among the first to adopt this new material. Acheson's company developed graphite liquid lubricants as well, using exfoliated synthetic graphite in oil. But, even with these niche uses and a clear patent case to support production in this instance, the combination of electrolytically produced graphite was too expensive to compete with mined natural graphite. Silicon carbide became a leading abrasive, and the Acheson process continues to be the foundation of the dominant production methods today. The International Union of Pure and Applied Chemistry recognizes Acheson Graphite as a type of synthetic graphite, but, as better methods have been developed since to create synthetic graphites, Acheson Graphite is an anachronistic name not used in anything other than a historical context. Today, graphene grown from silicon carbide is called *epitaxial* graphene.

Graphene layer growth from the decomposition of silicon carbide is now an extremely complicated process, in which the silicon is sublimed at high temperature as in the past but the atmosphere above the surface layer is variable. Tailoring the environment above the SiC surface allows researchers to produce graphene at better efficiencies than with an open air atmosphere. A 2009 *Nature Materials* editorial by Dr. Peter Sutter described an advance in epitaxial growth that involved removing air from above the silicon carbide surface and replacing it with an inert (nonreactive) noble gas atmosphere.[14] Since then, research has turned back toward reactive atmospheres. In a twist, three groups from across Germany devised a method where they glued a plastic made from many aromatic benzene hexagons onto a silicon carbide surface and found that this plastic actually drastically improved the size and quality of graphene monolayers produced from the silicon sublimation.[15] This work was inspired by another earlier paper, which fused CVD with epitaxial growth to improve the graphene yield.[16] It seems that somehow the combination of these two processes creates a product that is leagues better than either isolated method. If time tells that this combination turns out to be repeatable and economical, it could set the stage for graphene's everyday importance to skyrocket. What's more, it could even force out natural mined graphite from high-tech graphene uses. That could spell disaster for graphite mining companies who are betting their futures on selling to graphene consumers. This will be a development to keep close tabs on.

Expensive, rare, or otherwise valuable starting materials will generate significant demand for those starting materials, which would limit graphene's use in everyday materials. This would be a bad thing for everybody. After all, what kind of revolution occurs only for the superrich?[17] Therefore, it is absolutely imperative to find a way that graphene can be made reliably from a cheap (or how about free!?) resource. If graphene could be made from things that would otherwise go to waste, this would significantly decrease the long-term price of graphene so that anyone could have access to it.

If such a process were available, those who invented it would be regarded as highly as Fritz Haber, who won the Nobel Prize in Chemistry in 1918 "for the synthesis of ammonia from its elements."[18] Haber took nitrogen from the air and hydrogen from methane gas, combined them under high pressure and temperature over a metal catalyst to speed up the reaction, and boom! Ammonia came out of the reaction, ready to be put into fertilizer. Haber's invention quite literally feeds the world.

What starting material could we use for carbon as a feedstock that would not unduly tax typical sources of carbon, like fossil fuels or natural gas? Certainly, one option is to harvest carbon dioxide from the air and reduce it from CO_2 back to C. That is an extremely energy-intensive process, however, and no technological advances within the known laws of physics will reduce that energy demand.[19] That leads us back to thinking about something that is abundant, all around us, makes efficient use of capturing carbon, and can capture this carbon without direct energy input from humans. Plants. Plants take in passive solar light and carbon dioxide from the atmosphere and grow in most places of their own accord. Huge trees are carbon sinks made possible by photosynthesis. Lots of plant waste is generated per year, which might go toward creating graphene if it would otherwise take up space within a landfill. Invasive species of plants, like the rampant kudzu and bamboo in the southeastern United States, can serve as a feedstock, which has tangible *negative* impact on the local ecosystems. Turning invasive plants into graphene would be good both for graphene and for the environment.

James Tour took this to a logical extreme in 2011 on a bet. Tour had been thinking about the ways to use carbon already free around us in the environment, and had been successful in converting Plexiglas (polymethylmethacrylate)[20] to graphene, and table sugar was his next target. After having turned table sugar into pyrolysis-CVD graphene flakes on a piece of copper foil, one of his colleagues perked up, and dared Tour to make graphene out of six different carbon-based materials: cookies, chocolate, grass, polystyrene (Styrofoam), roaches, and dog feces.[21] This result is interesting, as the Australian lab mentioned above failed when using a copper foil substrate for their soybean oil conversion process. What these conflicting stories mean, however, is that there is vast room for improvement in our understanding of the way graphene forms from gaseous molecules. (If you must know, the cookies were Girl Scout cookies and the feces were from a dachshund.) Using the same method they employed with the table sugar, all of the unusual carbon sources produced small flakes of high-quality graphene. Tour and his coworkers stressed that no preparation or purification of these weird materials was necessary. In other words, the roach leg could be dropped on the foil, heated up, and come out as graphene. You can't even make a cake with that much ease. Tour's 2011 finding, combined with the "benzene glued on SiC" CVD-epitaxy findings from the German team in 2016, could provide a clear route to make large, cheap, defect-free graphene samples.

Graphene is composed of purely carbon as a single sheet in a flat hexagon pattern. You have heard this time and time again throughout the book before now. However, it absolutely bears repeating. Any changes to this structure mean that the resulting chemical is no longer technically graphene; it is a graphene derivative. To the layperson, this distinction may be abstract, silly, or unimportant, but the difference can make or break a product. In terms of defining engineering challenges, the difference is quite significant. Graphene behaves very differently from graphene oxide, and both behave differently from lithium-doped graphene. Take, for instance, the difference between two samples of exfoliated graphite from two different companies. One sample could have been exfoliated by a process that is rather harsh on the graphite, so that the exfoliation added defects of oxygen atoms or alcohol groups to the flakes. The second sample could have been exfoliated more gently, in a way that preserves the carbon-only structure without adding holes or tears in the flakes. Which is better than the other? How can you tell them apart? Both manufacturers slapped "Graphene" on the bottle and sold it to you at an exorbitant price; they must be indistinguishable in a product formulation and therefore you can just go with the cheaper option, right? Not so. The source of the graphene and how it was prepared has tremendous implications for its performance in a device. In chapter 2, we revealed that natural graphite did not have high enough crystallinity to allow Mildred Dresselhaus to determine graphite's band structure. In a similar vein, a batch of "natural" graphene with significant defects will degrade any application sensitive enough to require pristine graphene. A device might not work at all, or may just work worse than expected.

Standards do not exist yet for graphene production, and not all companies are on board with establishing standards at all. These standards could take many possible forms and do not necessarily mean legal regulation. That would be quite obviously an extreme measure, and would be unenforceable in other countries. Considering the international playing field for graphene, this would be a significant hindrance. Nobody wants that. However, at this point in the game, most products labeled "graphene" on the market are not actually graphene. Rather, they are thin flakes of graphite that can be up to a few hundred layers thick. Some manufacturers are able to produce flakes with a high yield of monolayer graphene, and these companies will gladly tell you that they produce a guaranteed percentage of monolayer graphene, with most of the rest of the sample consisting of flake aggregates between two and ten layers thick. A word to those of you who are interested in using true graphene for an application—ask about these flake thicknesses from your supplier. It is absolutely critical to take what they say to an independent lab for verification to establish a definitive level of trust.

Ideally, standards set forth should grade graphene taking into account parameters like the yield of monolayer flakes, the size of those flakes, and the elemental analysis of the sample (at a minimum). That way, a vendor can stand behind the production cost of their so-called graphene sample, rather than jacking up the cost for some graphite that has been pulverized in a kitchen blender. *Caveat emptor*. On the other hand, if a vendor is selling high-surface-area epitaxially grown graphene with a repeatable or verifiable certificate of analysis, then you may have a justification to pay more for that sample. Will blender graphite always perform worse than pristine Chemical Vapor Deposited or epitaxially grown graphene? That is a question for your application engineers to determine. It is important for inventors to recognize that incorporating graphene or graphite into a formula will not simply change the material to be very graphene-like. As often as it is touted as a miracle material, we must not treat graphene as a modern day alchemical wonder. These mixtures or composites are more complex than that, which means they require their own due diligence.

The next ten years will see a proliferation of graphene-enhanced products in the marketplace, but we are only at the very infancy of the lifecycle for market viability. We have been fortunate that an infancy has even begun. According to Zina Jarrahi Cinker, executive director of the US-based National Graphene Association, graphene entrepreneurship almost perished before it had a chance to mature as a technology. In part, excitement generated by the 2010 Nobel Prize and the subsequent articles that sang graphene's superlative praises set expectations unrealistically high. Investors raced to pump money into early startups, but hoverboards and flying cars failed to materialize. This burned many venture capitalists on graphene for a while, and even the US government decreased the number of Small Business Innovation Research/Small Business Technology Transfer (SBIR/STTR) grants that it gave out, following lackluster research and development projects. That attitude is starting to come around. Investment in graphene is not merely limited to the United States, United Kingdom, or Singapore. Ventures have been launched all across the globe, and funds are not restricted within national borders. As a symbol of international cooperation, China's president Xi Jinping visited the National Graphene Institute in Manchester, in the United Kingdom, following a partnership deal between the institute and the Chinese company Huawei.

Graphene's excitement generates funding for research, which can translate those results into products. Not all of that excitement is beneficial to graphene's outlook, though. Most of the articles on graphene focus on the gee-whiz "material of tomorrow" aspect that generate well-deserved excitement for research progress. Far too few science articles gain widespread popular attention as it is, but sometimes coverage of a paper can accidentally misrepresent findings in search of a catchy headline. We (the authors) sympathize with that plight; it is not easy to come up with a book title either! Entrepreneurs want to meet or beat that excitement to generate investments. That desire is completely natural and understandable. However, creating a sustainable business model requires a bit of measured realism. A team would be wise to use a certain level of restraint in their pitch decks to investors. More than that, though, the team needs to not only understand the differences between different graphene grades, they need to be able to communicate those basics to investors. That way, everyone can maintain even expectations across the board.

Product development lifetimes are almost never overnight sensations, and it would be naïve to believe that graphene will be any different. Another technology has recently reached commercial maturity and can be used by young companies as a roadmap to their own financial success. The engineer Henry J. Round did not likely envision a successful application when he accidentally created the Light Emitting Diode (or LED) in 1907. He was experimenting with a sample of carborundum when he noticed the sample glowed yellow—but not because it was extremely hot.[22] A hundred years later, companies released the first consumer LED lightbulbs. Unfortunately, the first bulbs were not (in retrospect) very good. While they were noticeably cooler than incandescent bulbs and had better advertised lifetimes than compact fluorescents, color management and light intensity were hard to judge. Who knew how many lumens (light intensity units) they needed to read a book by the bedside or illuminate a full thanksgiving dinner? These units could be confusing, at first. Most importantly, though, early LED bulbs were *extremely* expensive. A single bulb could cost well over thirty dollars. The package could say all it wanted about how each bulb would pay for itself seventeen times over that in the cost savings of electricity during the bulb's lifetime. But would the bulb last as long as the advertised twenty-five years? That's a pretty significant commitment to an unproven technology. Don't forget that customers' houses were already lit by incandescent bulbs by the time Round was observing the first *electroluminescence* experiments. The market got extremely lucky this time around. Enough people bought those pricey first-round bulbs for competition to grow, and the market produced better bulbs. They became more energy efficient, and the price per bulb plummeted. While modern LED bulbs are still more expensive than incandescent bulbs, they "pay for themselves" on much shorter time scales than before. Thus, customers are seeing an ever increasing benefit to market maturity as time passes. By waiting for graphene to be better understood before bringing products to market, the investment costs of a company's R&D can be decreased, allowing the consumer's costs to decrease along with it.

LED lights can also serve as an illustrative example of competitors coming together around shared values to create a standard platform so that customers can make an informed decision. Putting lumen output on the box was for nerds. It would never work for a new customer base because there was no familiar comparison for them to turn to. Incandescent bulbs, on the other hand, had settled on using power consumption as the standard for relative light output. Anyone who has purchased a lightbulb knows roughly what a sixty-watt light bulb should look like. The problem is, LED bulbs are more efficient. With the familiar units that the customer holds, trying to sell a five-watt bulb would only end up in disaster. Therefore, LED bulb manufacturers arrived at an interesting temporary compromise—the bulbs would be advertised as watt-equivalents. A sixty watt-equivalent LED bulb would be just as bright as a true sixty watt incandescent bulb but use much less energy than that in practice. Now that customers are more familiar with LED bulb technology, manufacturers and retailers are beginning to educate customers to be comfortable in both lumen and watt-equivalent unit expressions. This is a sure sign that retailers eventually want to phase out the watt-equivalent phrasing and move to lumens entirely.

Standards within graphene production should be agreed upon in a similar light. Manufacturers should work together to create a consensus that is fair and profitable for most, if not all, of the industry players. Zina Cinker and the National Graphene Association are working with leading companies and nongovernmental organizations to try to put this framework into place. Hopefully the market sees another success story. From there, research will continue on in production methods so that by the time graphene nanoplatelets have reached industrial maturity the first macroscopic (>1 cm^2 or >0.25 in^2) pristine sheets will be nearing fruition. The graphene nanoplatelets are great and all, but there is far more room for innovation available in large-area applications.

Graphene is so thin and light that storing the material for later use is problematic. With an atomically thin material, the slightest disturbance brings with it the possibility of the graphene sheet being altered, radically changing the properties of graphene and even rendering the material useless. To allow for transportation, graphene sheets are often stored under water or in abrasive solvents, requiring a series of intricate preparation steps before use.

Isolating graphene flakes from bulk graphite is a process that yields sheets that are isolated from their natural surroundings, wrested from the conditions under they were formed deep beneath the Earth's surface. Due to this fact, graphene flakes isolated from natural sources have uneven and random shapes, and no particular pattern or distribution emerges. While it is true that some sources of graphite do yield larger flakes than others on average, the fact still remains that any industrially useful material will require tight quality control and standards by which to create useful products.

This is especially true in the realm of creating graphene-based semiconductors, as it has been found that the width of a sheet has a profound effect on its electrical properties. Conductors, semiconductors, and insulators are divided from one another based on their ability to shift around electrons in the presence or absence of extra driving forces. The energy of electron orbitals in conductors like metals have a small (if any) gap between their lowest energy static state (called the valence level or valence band), and the higher energy state where they can move about (called the conduction level or conduction band). Little to no energy needs to be applied in order to get these electrons to move around in a sample. In semiconductors, there is a small but critical gap in between the valence and conduction bands.[23] This gap means that energy needs to be put into the material by heat, photons, or electric potential in order to make electrons flow. From this gap, there is a definitive state when the material is "off," when no external influence exists, and there is a definitive "on" state, where the electrons flow within the applied influence. Insulators have a very large gap between the valence and conduction bands, which effectively means that no useful current will flow with any applied outside energy.

Researchers found that if you have a long piece of graphene but it is very narrow, the sheet behavior changes. They have been able to determine that there is a subtle gap between the highest energy electrons and vacant energy levels, and it is only within the vacant energy levels that the electrons can flow around freely. This subtle gap opens up as the graphene material becomes more "molecule-like," and the electron bands become more like discrete electron levels. When this gap doesn't exist, the typical graphene behavior emerges—the sheet conducts like a metal and electrons move freely around the sheet without a problem. There is no on/off dichotomy, instead there is just the pure conductor that graphene is hailed as for the moment. Introducing the gap, however, also introduces the on/off dichotomy necessary for graphene to become a useful logic unit. Logic units underpin the function of computers, and making them from graphene could make them much more energy efficient. This would grant cell phones better battery life and make your laptop cooler on your legs.

At the moment, NASA is researching ways to process waste carbon dioxide (CO_2) from astronauts' breath on the ISS (International Space Station) into graphene. This improvement to the life-support system would have a twofold bonus. For one, a waste material like CO_2 otherwise requires sequestration with special chemicals that need to be shipped up with special deliveries from Earth. Processing the CO_2 into graphene would mean fewer resupply missions would be necessary. As fewer resupply missions would be required to maintain the station, operation and maintenance of the station becomes cheaper. Turning CO_2 into graphene provides another benefit as well; the resulting graphene could be incorporated into new solar cells, or could be put to use in the water purification systems, or a thousand other possibilities, rather than trying to eject it out the airlock. This possibility helps to lengthen the umbilical cord between the ISS and Earth. Eventually we need to cut that umbilical entirely, if we are to ever send humans on extended missions to other planets and beyond.

Luckily, there is a side benefit for us Earthlings as well. A process like this would also be able to take CO_2 from the atmosphere and turn our own breath into organic electronics or a million other things that graphene could find uses in. While turning CO_2 into graphene would not be cost effective nor energy efficient on Earth (right now), abundant power from solar cells aboard the ISS would provide the kick necessary to strip oxygen from the CO_2. Companies could "mine" the atmosphere to take carbon dioxide from processes that can't help but produce it, and turn the waste gas into a raw material for further products. The "waste not, want not" principle that every hiker and explorer knows well means that a system designed for reuse will ultimately increase the chances of a mission's success (whether it be on Earth or in space), while also minimizing environmental impact. Redundancy on Earth can only be a good thing. In outer space, it is an absolute requirement.

This is not a new concept, of course. Turning effluents from one process into raw materials for another transforms the idea of waste. This continuously regenerating cycle is the cornerstone idea behind William McDonough and Michael Braungart's Cradle to Cradle philosophy. Companies can profit off of what they would otherwise have to pay to throw away, and they can revolutionize their image if the trash also finds itself as a consumer benefit.

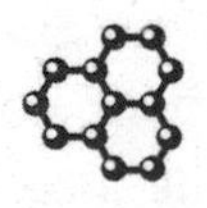

But what are the potential hazards of using graphene on an industrial scale? Is there nothing about graphene that is potentially dangerous?

With all the wonder and awe behind the graphene revolution, we know dangerously little about the potential side effects or dangers of graphene. The medical research about graphene is rather sparse when compared to the extremely thorough treatment it has received in the hands of physicists and chemists. We have little idea about what having tremendous amounts of graphene produced each year could do to our bodies, the environment, or to other living things.

A 2016 review article by Lingling Ou and other researchers summarized the state of medical toxicology research on graphene. The most poignant line in the conclusion states, "Many experiments have shown that GFNs [graphene-family nanoparticles] have toxic side effects in many biological applications, but the in-depth study of toxicity mechanisms is urgently needed."[24] This sentiment is peppered throughout the paper, in the discussions of health studies that have been performed on different cell culture lines in order to determine the toxicity of graphene on different parts of the body. The research group concluded that the health effects of nanoscale graphene and graphene-related flakes (such as graphene oxide) are enough to raise a concern but that the results are not mature enough to understand how these particles affect cells. An aspiring young researcher would find an abundance of opportunities to carve out their own research niche in this area.

As graphene flakes available for research at the moment are only a few nanometers or micrometers on a side, all of the modern research into toxicity has focused on the effect of graphene and graphene oxide flakes at this small scale. Cells and viruses are prevalent at this length scale, and it is crucially important to figure out what will happen to graphene that is incorporated into products that we will consume and use. It would be a tremendous tragedy to make a consumer product with so many wonderful benefits, yet discover that normal wear and tear on the item renders it harmful or deadly. Free-floating graphene is not a naturally occurring substance for most living things.

Some limited early evidence suggests that pure graphene might not be good for cells, and here's why: You may recall that cells are globules of lipid membranes that surround an inner working of other smaller cellular machinery. The fat membrane also contains a number of proteins; it is these proteins that allow the exchange of nutrients and waste between the cell's membrane and the environment. Since proteins' functions are directly tied to their structure, it is important that nothing disturb the way that these proteins fold and unfold. Proteins' structures often involve interactions between atoms that are not directly attached by the shared-two-electrons covalent bonding. Instead, medium-strength electric forces (called dipoles) based on the arrangement of atoms in space cause a protein's amino acid chains to twist and fold into shapes characteristic of the protein. Since these dipolar forces are able to form and break with relative ease, the proteins are particularly sensitive to other molecules that form these types of interactions. Graphene and graphene oxide form these dipolar interactions, and the flexibility of the sheets means that they can twist to conform to the outside of a protein. Once this happens, the protein is in severe danger of being pulled apart and misshapen.

Basically what would happen is that the parts of a protein damaged by graphene would stick to the surface of the sheet. This would, in turn, disturb any other interaction that the protein would have normally, causing the protein to lose its structure and become disabled (biologists and biochemists would call such a protein *denatured*). When you disable a protein, it is no longer able to perform the function it was designed to do, and this could cause all sorts of internal problems for the cell. Cells need to breathe and shuttle molecules all about their internal structures; they need to take in nutrients and get rid of waste. This all happens very efficiently because specialized molecules do their job very well. When an interloper like graphene enters the cell, and it cannot be dealt with by the cell's normal functions, this causes incredible trouble for the cell. One reaction to graphene entering a cell could be apoptosis, which can occur when the mitochondria within a cell become overly stressed. This stress cascades into a chain reaction, until the cell bursts apart and spills its internal contents into the surrounding environment. It's sort of like a cellular supernova. The trouble is, this won't destroy the graphene sheet, and the graphene has just been released back into the environment, where it poses a danger to other cells.

Graphene is not only a danger to proteins, though. It also poses a hazard to DNA and RNA within the cell. As graphene is only one atom thick, it is able to slide in between the stacked base pairs of nucleic acids, disrupting the helical structure of the chromosome. We already see effects like this from benzene and the other aromatic hydrocarbons, and this is the fundamental reason for the toxic nature of polycyclic aromatic hydrocarbons, the family of molecules that are basically graphene at extremely small sizes. This interruption would lead to possible transcription errors, causing mutant cells to form. While having the X-Men come about as a part of the graphene revolution sounds pretty cool, this is still likely more in the realm of science fiction rather than science possibility. Sorry, Stan Lee, it is more likely that any mutations would not be beneficial at all, and perhaps harmful.

Fortunately for us humans (and all higher life-forms, in fact), we have a nucleus that contains and protects the genes from a lot of harmful materials. Graphene has the potential to cross the nucleus barrier, but the chances are highly dependent on the graphene-based system in question. This fact is beneficial for complex multicellular organisms like us, as compared to bacteria, which do not have a nucleus to contain their DNA. However, our DNA is still susceptible to damage when the cell replicates. When the process of mitosis begins, the nuclear membrane breaks down, and the chromosomes are exposed to the rest of the internal cellular environment. If graphene flakes are present within the cell itself, they can insert themselves into the genes and pass on mutations to the next cellular generation. A study performed on mice models showed that graphene injected into the blood was more than twice as mutagenic as cyclophosphamide, a common benchmark chemical.[25] Graphene and graphene oxide have different toxicities, largely due to the difference in chemical properties that allow them to cross the cell membrane at different speeds.

That said, this toxicity to cells largely focuses on what happens when the flakes cross the lipid membrane surrounding the cell, and end up inside. There needs to be much more research to figure out how the graphene structure affects each different cellular organelle and what happens during each step of the cell death. Studies on graphene toxicity have shown that it can definitely cause complications within the body; however, we are still only beginning to learn the finer details about what happens and why. We absolutely do not want to release this new material on the world, only to learn that it is another persistent toxin or pollutant.

The toxicity of graphene nano-sized flakes is not a concern at this point. Our hesitation should come from the lack of full understanding behind graphene's action on the body. But should we be concerned about larger sheets, once manufacturing companies are able to produce swatches of graphene that we can pick up and handle with our hands? It is important to note that answers to this question fall under speculation by the authors. This question has not been addressed within the realm of current medical science.

The large graphene sheets would be unable to enter the cell, and therefore many of the toxic properties already exhibited by common graphene flakes would no longer be a concern. The danger to cells would not go away just because a sheet is larger than a cell itself. The danger could instead be to a whole group of cells at once, causing destruction to whole swaths of skin, lung, blood, or other tissue's cells. A sheet adrift in the wind, if inhaled, could lodge in the lungs and block airflow to localized areas of the lung. This is the danger of the sheet being so thin and flexible; the total volume of an atomically thin material is extremely low and can fit into tight spaces if it gets bunched up. Another possibility is that aggregates of graphene sheets could clog capillaries, veins, or arteries. Without blood flow, tissues would die. Our cells still have proteins and other organelles on the cell surface, and just as small graphene flakes adsorb onto the surface of cells to bind with those surface proteins, so too can a macro-sized sheet. If a transport protein, say for the sodium ion, were affected, then the cell would be unable to regulate how much sodium comes into or goes out of the cell, leading to a dangerous electrolyte imbalance. If a recognition protein were affected, the cell would effectively become blind, as it would be no longer able to recognize its environment. This last scenario would be especially detrimental to the immune system, where white blood cells need to be able to recognize pathogens in order to kill them.

Despite these concerns, particularly the concerns regarding the incompleteness of the information, some researchers have found promising early results. In 2013, Professor Alexander Star coauthored a review article outlining the latest developments in carbon nanotube degradation within the body.[26] While we have described differences in electronic and physical properties of carbon nanotubes and graphene in earlier sections, it is well within scientific possibility that the biodegradation of carbon nanotubes and "internal" graphene atoms would proceed along similar pathways. Once a nanotube or fullerene is broken into different pieces by a chemical reaction, the pieces have unstable edges particularly vulnerable to further attack.

This functions somewhat like starship shields in science fiction. The ships are vulnerable to high-energy damage from asteroids and superweapons, but traditional laser weapons are blocked. If you destroy the shield generators (by exposing unstable dangling edges), however, then the ship as a whole becomes vulnerable to destruction through subsequent damage from a broad spectrum of different sources. Remember that the edges of graphene are less stable than the center of the sheet, which means that chemical modifications to a graphene flake are easily accomplished at the edges. This is not to say, however, that central or non-edge carbons within the graphene sheet are impervious to chemical modification, though. Oxygen adds to graphene flakes to produce graphene oxide, which reacts differently within cells.

Hydrogen peroxide attacks graphene and other carbon nanomaterials, assisted by an enzyme called a *peroxidase*. Peroxidase enzymes are found in many different living systems, and the enzymes assist in degrading harmful chemicals within a cell by attacking these chemicals with hydrogen peroxide.[27] The humble horseradish, a highly underrated root, contains an enzyme called horseradish peroxidase within it that has shown an ability to attack and degrade a great number of different organic compounds. This peroxidase is used in wastewater treatment plants, in fact, to destroy harmful chemicals within our municipal water systems.

The horseradish root proved to be an early quality-control measure when a French pharmacist Louis-Antoine Planche discovered that fresh horseradish placed in a solution of resin from the guaiacum tree rather quickly turned a blue color. Planche was working on ways to detect guaiacum adulteration of another product, jalap resin, which he was importing. This allowed him to spot batches of his herbal remedy that had been tampered with by unscrupulous suppliers. Unbeknownst to him at the time, it was the peroxidase enzyme in the horseradish root that enabled him to detect the fouling ingredients. Interestingly, the guaiacum colorant was eventually adopted as a clinical diagnostic tool assisting in the detection of non-visible blood in stool samples. Peroxidases from enzymes in blood would react with a paper strip and oxidize the colorless acid into a bright blue compound, in the same way Planche's horseradish had worked.

As horseradishes are plentiful, and the biochemistry behind horseradish peroxidase is especially well-understood, horseradish peroxidase has become a model enzyme for testing out the biodegradability of many different types of nanoparticles *in vitro*, or outside of the body. Star noted in his review that only nanotubes with initial defects were affected by the horseradish peroxidase; no defect-free nanotubes were degraded. The shield must be deactivated, if any attack is to be attempted. These enzymes are important for their roles in regulating the breakdown of graphene and nanomaterials that will eventually end up in our drinking water, our gardens, and ultimately, our food.

When it comes to the ability of our bodies to deal with graphene and carbon nanotubes, our first line of defense is the same as that deployed against bacterial invaders. White blood cells will undoubtedly encounter graphene flakes within the bloodstream, so it will be important to know if and how these cells will deal with the potential threat. Star and his coworkers were able to determine that an enzyme called *human myeloperoxidase* (hMPO) was able to degrade carbon nanotubes *in vitro* as well. After a white blood cell takes in a bacterium, the cell releases hMPO. The enzyme then works to break down the bacterium's cell wall and kill it. Star theorizes that the hMPO degrades carbon nanotubes by creating an acid capable of creating defects in the nanotube walls, thereby creating the very first chink in the shield. While breaking down carbon nanotubes may only lead to the creation of graphene or graphene oxide flakes, it is one step in the ultimate chain of custody which all of these nanomaterials manufacturers will be responsible for, should they desire to maintain a place in proper stewardship of our environment. We must understand how nanomaterials interact with our anatomy to discover how to best take advantage of their useful properties without accidentally making persistent poisons. For example, graphene oxide, just as the defective carbon nanotubes, is biodegradable, but pristine graphene may require prior oxidation to graphene oxide before our bodies will be able to handle it.

As we gain the ability to specifically tune or manufacture graphene flakes to custom size requirements, we must look again at carbon nanotube research. "Long fibers and large aggregates of CNTs," Star writes, "which are difficult for [cells to absorb], typically induce asbestos-like [symptoms]."[28] It doesn't take a medical researcher to realize that having a new asbestos scare on our hands would be disastrous for a material that offers such considerable promise. Graphene nanoribbons, carbon whiskers, and carbon fibers could all cause bodily harm if their tangles and twists cannot be properly disposed of by our lymphatic system. A Miner's Lung for the modern age should not be named the Graphene Liver. Laborers in future production facilities should not have to worry that their work will destroy their body.

As a part of the University of Manchester's Graphene NOWNANO program, Drs. Kostas Kostarelos, Cyrill Bussy, and Sarah Haigh are collaborating across departments and disciplines to research the mechanisms underlying biodegradation of graphene and related materials within the body.[29] They specify that graphene-related materials are a part of their research repertoire because tailoring graphene to biological applications will require adding molecules and functionality to graphene. As we have tried to emphasize in this chapter, these additions would no longer allow the graphene to be designated as pristine graphene itself. It would not be technically correct to call a modified graphene superstructure graphene; that would be misleading. And, as we all know, technically correct is the best kind of correct.

What if an enzyme or other mechanism within cancer cells (but which does not exist within normal healthy cells) is able to provide that first defect to start the chain reaction? If cancer cells had a reaction in the cell that regular cells do not, chemotherapy drugs could be very exactly delivered to cancerous sites without damaging healthy cells. The carbon-nanotube-encapsulated chemotherapeutics could be delivered intravenously. Normal cells wouldn't uptake (or, absorb) the CNTs in large amount. Even if they were absorbed, the normal cells would not break down the CNT walls and after apoptosis the drug would be free to travel around the body again. Only when the system encountered a cancerous cell, was uptaken, became oxidized, and finally degraded, would the drug spill out and kill the cell. Fullerene components and graphene flakes would already be oxidized from this local environment, which would mean any of this material that escaped into the surrounding tissue could be handled by the normal mechanisms.

Graphene's potential to change the course of innumerable industries is only limited by the imagination and cunning of business leaders who can share a common vision alongside a knowledgeable chemist, engineer, or physicist. Bolder, more enterprising technologies will develop by adding different molecules to graphene, treating it as a scaffold onto which biomolecules can be grafted. This would make the ultimate nano-cyborg—living or life-adjacent structures atop a graphene surface may sound like fanciful science fiction now, but passive sensors for chemical and biological weapons will need to increase in complexity to match the pace of development of those weapons. A complex sensor could, in theory, contain an array of proteins selective for gaseous chemicals. If a weapon chemical were present, the protein would bind to the weapon and undergo a change. From there, an electrical or magnetic signal would be tripped in the graphene sheet, alerting a computer to the weapon's presence. Specially engineered molecules like proteins or nucleic acids could bind these weapon targets without error and might never need replacing if they are designed to be "rechargeable."

Graphene as a coating material could even change industries in the short term. Since graphene is mostly nonreactive and very hydrophobic, any surface coated in a layer of graphene would move through water with decreased friction from water-metal surface tension. A graphene layer on tanker ships would make worldwide shipping more effective. Adding a graphene layer onto a windshield would create a surface that was not only transparent (because graphene itself is transparent) but would naturally repel water and increase driver safety in rainstorms. Want to reduce air drag on a high-performance car? Ensure that its shell is perfectly atomically flat by encasing it in graphene. Maybe an especially talented engineer in the future will design a vehicle with perfectly laminar (i.e., smooth and regular) flow over the car's body, eking out a few more horsepower from the engine and a few more miles per gallon from the tank. In the upcoming chapters, we'll address some of the visions for inventions that are further afield, looking toward the time when large-area graphene wafers are available.

Chapter 5

COMING SOON TO A STORE NEAR YOU? *OR*, SO WHAT?

Up until now we've been learning about graphene and how it is unique because of its incredible physical and electrical properties. We've learned about how it was accidentally discovered and why there is so much controversy surrounding that discovery among some researchers. We knew that it was hiding in plain sight, but we never were quite able to convince ourselves that it would be particularly stable if it were isolated as a single lonesome sheet. It was so well hidden, in fact, that several related but still distinct allotropes (the fullerenes and nanotubes) were isolated and characterized first.

We have even seen graphene make its way into a couple of interesting applications here in the recent years. It's particularly difficult to flip through the Science section of a newspaper or technology magazine without coming across new studies espousing the wonders of this material. Some of them come across as downright science fiction. But for all the hype, what are we really going to see out of our investments in this special carbon? Is there really a miracle that we can expect to come from all of these fancy words and extremely complicated experiments? Or is this some dumb pipedream—the hardware equivalent of software's "vaporware" that promises big but never delivers? When are we going to actually see a product on our shelves—one that we can buy and feel confident will work as advertised? Science has made a lot of promises. When are they going to pay up?

Soon.

This is a particularly exciting time of innovation. The two primary properties that make graphene especially valuable—its strength and its electrical conductivity—are going to see the most number of direct consumer applications. Its strength will be involved with many safety-inspired or construction materials. Its electrical conductivity will allow us to passively capture energy from our environment and charge small specialized circuits with that power. We'll be able to see interesting new applications of "smart shoes" or "perpetual wristwatches" powered from body heat.

At the risk of overhyping this revolutionary material, this chapter will explore the vast "what if" potential graphene offers—both now and in the future. Keep in mind that not many "super materials" discovered in the last few decades have lived up to the hype surrounding them, but it is beginning to appear that graphene will succeed where others failed.

Let's assume we're building a home in the community of Anywhere, USA. Like most places in North America, Anywhere is challenged by extreme weather conditions: blizzards and high winds in the winter, tornados in the spring and fall, hurricanes in the summer, and earthquakes just about any time. All in all, there are many ways Mother Nature can damage or destroy our new home, and we want to make it as resilient as possible within our budget.

After consulting with an architect and settling on an overall design and floorplan, we need to consider its *foundation*. Many homes in Anywhere sit on clay, with all the moisture retention problems that entails. Given that we are in a region rich in tornados, our new home should have a basement—making the necessity of keeping ground moisture from seeping into the basement a priority. For this we select a poured concrete floor and concrete block walls. On the outside of the walls we're going to paint graphene-enhanced paint that will stop water seepage completely. In addition, the waterproofing paint will act as a barrier to general environmental degradation and provide additional strength to the structure.

Within the basement we're going to install a *tornado shelter*. In 2011, a series of tornados swept through Alabama, killing three hundred people. Just days later a major tornado swept through Joplin, Missouri, killing hundreds. These events make you take notice and consider the future safety of your family. Similar events, like hurricanes and strong storms, impact the East Coast all the time. We need to prepare ahead of time for the worst-case scenarios.

Graphene, being the strongest material ever measured by scientists, is perfect for use in construction of our shelter.[1] Its intrinsic strength, the maximum stress that a defect-free material can withstand before breaking (having all the molecular bonds pulled apart at the same time), makes it ideal. According to James Hone of Columbia University, one of the scientists who measured graphene's intrinsic strength, in an interview with *Physics World*, "To put things in perspective: if a sheet of cling film were to have the same strength as pristine graphene, it would require a force of over 20,000 Newtons to puncture it with a pencil. That is the force exerted by a mass of 2000 kilograms, or a large car!"[2] Given that many injuries or deaths during a tornado or hurricane are caused by flying debris, this is the kind of protective coating we would like to have on our shelter.

Next comes the *framing*. We're going to want the frame to be as strong as possible in light of the tornado, hurricane, or earthquake risks we're facing. For the very same reasons we are choosing to strengthen our tornado shelter with graphene-enhanced materials, we will be similarly strengthening the framing that keeps our future house standing.

At this point, we start thinking about the utilities. It turns out that graphene is an excellent conductor. In fact, and as we'll discuss later when we start talking about the items we will put into the house, it has other useful and very interesting electrical properties. For now, we are concerned about just piping electricity into the house as efficiently and affordably as possible. We'll begin by looking at the solar panels that will be installed on the roof.

Instead of today's silicon- or germanium-based solar cells, our rooftop array will use—you guessed it—graphene-based solar cells. Graphene is not only more efficient at producing electricity (releasing multiple electrons per incident photon instead of just one), it works across a wider part of the electromagnetic spectrum, allowing previously unusable light from the sun to produce electricity instead of being reflected or absorbed and turned into heat. This unusable light could cause damaging heat—which is why modern solar cells need to be cooled in order to operate efficiently. Graphene gets around this by actually using this light to release electrons. These graphene solar cells are extremely lightweight and flexible, meaning that we don't have to limit the solar cells to the roof. Graphene photovoltaic cells can be attached to any sun-facing surface on the house, including the south wall, which would generate peak power in the winter, at exactly the time it would be most needed in the utility cycle.

We might even go one step further and buy graphene-solar-cell-covered windows, which have embedded below them a thin layer of liquid crystals that allows us to use the power generated by the window covering to provide at-will dimming of the natural light. If we want complete darkness in our bedroom on an otherwise bright and sunny day, we can simply vary the current flow from the window's photovoltaics through the liquid crystals to block out the incident light.

Given that many families these days are seldom home and using electricity during the day, when solar power is useful and most easily generated, we will also equip the house with a graphene energy storage system to save as much unused solar-generated power as possible for use when we need it: at night. For this, we will turn to supercapacitors. Unlike a traditional battery, which stores electric energy using strictly chemical processes, a supercapacitor stores electrical charge on the surface of electrodes—an effect similar to what you experience when you rub your feet on carpet and generate static electricity. Non-graphene supercapacitors already exist, but they are limited in the amount of charge they can store before breaking down. With graphene, the energy storage density of a supercapacitor can be as good as or better than a traditional chemical battery—at a much lower cost, smaller size, and lower mass.

Our efficiently provided electricity will then be sent to the ultra-high-efficiency heating system that uses graphene heating elements. UK-based

Xefro is building a system that they estimate will reduce home heating costs by 25 percent to 70 percent.[3] Xefro uses a graphene ink to make the heating elements and reduce the number of heat transfer methods of getting the generated heat dispersed into the rooms in which it is needed. Our new house will have wireless connect controls that will activate room specific graphene heating elements embedded in the floors to produce heat only when the room is occupied. The large area in which the heat is produced (the entire area of the floor) will allow the room to heat up quickly and only when needed. Other rooms not in use can be kept at much lower temperatures, saving electricity and money.

The next utility to be installed is *water*. The recent crises in Flint, Michigan, and elsewhere highlight within the industrialized world a problem that has been facing the developing world for centuries—the need for clean water. Due to a series of poor decisions and bad luck, the citizens of this American city have been exposed to lead-contaminated water for an extended period of time, and the solutions proposed to fix the problem rely on repairing or replacing hundreds of miles of water pipes throughout the community. Such massive infrastructure projects take lots of time and money to complete, forcing the consumption of bottled water in the interim.

In our new home, we will install simple graphene oxide membranes to filter all water contaminants, not just potentially offending lead. The graphene-based membranes to be installed are designed to remove heavy metals, organic toxins, and pesticides (as well as other common contaminants) with near-perfect efficiency.

And why not put these graphene filters at the other end, so to speak, and filter the gray water that would otherwise leave our house and flow into the city's sewage system? Such a filter could allow cleaned water to flow back into the house's potable water system, contaminant-free for reuse, allowing only the most contaminated of the waste sludge to flow into the sewers for more rigorous purification and disposal.

For efficiency and uniformity, we next plan to install graphene-based flexible lighting strips on the ceilings and walls of every room in the house.

Instead of the traditional light fixture or lamp, each containing a bulb to produce a discrete source of light, thin, lightweight, and transparent graphene-augmented strips will be applied and connected to the house's electrical power system. This is a matter of personal preference: some like a bright, uniformly-lit room without shadowed corners. By having the light emitted from everywhere, or, perhaps better stated, from a non-discrete source, we can make sure there are no dark spots in the room, and it won't matter where the light is located relative to whatever we are viewing.

But wait, the construction of our new home isn't the only place where we'll be using graphene-enhanced products. We're through making decisions regarding the construction of our new home, and the details, like actually constructing the house, are now in the hands of our capable and competent general contractor. Let's now assume that it's Saturday and time to take care of the usual family business—running errands with the family.

Our first stop is the local pharmacy, where we need to pick up some items for our medicine cabinet. Actually, we just need to restock the cabinet with some adhesive bandages. A parent can never be too careful, and the stock at home is running low since the family's recreational sports activities really started up. There are the typical store-brand bandages with plain-old cotton swatches on them. But what's this new brand here? Antimicrobial graphene bandages? The box claims that they not only fight infection but prevent it entirely by keeping the bacteria from growing in the first place! Any time one of the bacterial cells approaches the vicinity of a graphene sheet, it's promptly sliced apart. (Remember the water filtering properties of graphene mentioned above? If you think of bacteria as a contaminant, then you'll understand how the graphene sheet can "filter" it out.) Once the bacteria is filtered, sliced, and diced, your body can easily take care of disposing the rest. This will help keep the bacteria cells from dividing out of control and keep exposure to a level that your kids' bodies can manage by themselves.

So now we don't have to buy both the bandages *and* the antimicrobial ointment? That sounds like an excellent money-saving idea, even if the bandages themselves are just slightly more expensive. You probably heard of these bandages being used at hospitals in the area, and they were especially popular for wound care that would usually require antibacterial creams because the graphene bandages circumvent bacteria's ability to evolve a resistance to the creams. In fact, a recent column in the paper interviewed the founder of this company, praising him for his role in reducing hospital deaths due to infections from antibiotic resistant strains of so-called "superbugs." What's an extra dollar or two to ensure that our family is even safer and to prevent the proliferation of superbugs at the same time? It's a win-win for us and the community.

On our way to the next stop, we receive a call from the doctor. Our son's x-ray results show that his soccer injury didn't result in a broken bone, just a bad sprain. What we don't know is that the x-ray machine they used to make this assessment doesn't work the same way as the ones used when we were younger. Instead, the machine uses graphene's 2-D structure to produce plasmons (surface waves), which in turn trigger a finely tuned, highly targeted pulse of x-rays, with far less leakage than previous x-ray machines, exposing our son to far less x-ray radiation than was possible with previous x-ray machines.[4] His sprain's cast, instead of being made with heavy and unwieldy plaster, will be made with a thinner graphene-enhanced rubber composite. The increased support and reduced hindrance from this special mixture will reduce his recovery time so that he will soon be back on the field where he belongs.

And thinking about sports, we decide to stop by the big-box sports store. There's a big holiday weekend coming up, and the family wants to go hiking. Our mountain-fanatic friends recently picked up these new socks that they're just raving about. Supposedly, because of graphene within the silk fibers, they are extra smooth and will keep our feet from stinking even after a long day on the trails. They work similarly to the bandages that we just picked up. Shirts and pants made from the fiber keep thorns from scratching us or ripping the material. It's so soft and smooth, in fact, that it reduces chafing from extended wear on a hard day. It'll be nice to not stink so much after mowing this summer. Body odor is caused by bacteria, and you know what happens to bacteria that try to pass through graphene . . .

But what's that on aisle three? These new bikes are sporting not only carbon-fiber frames for reduced weight, but their tires are even molded containing graphene in the rubber. We think, *Surely, this must be a gimmick.* But, as often happens, curiosity gets the better of us and we attract the attention of a nearby associate. "What's with these tires?" we ask.

"Oh yeah, they're spectacular. I have friends who are seriously hardcore mountain bikers, and the graphene flakes in the rubber really increase grip on the trail and help the tire last even longer.[5] The dude who invented this must've been a genius. They'll last practically forever, compared to regular tires." It seems that the bike helmets, too, have hooked onto the graphene craze. They claim better energy dissipation for reduced impact to the skull, which means a safer fall in the case of an accident. Bike frames made from a graphene composite will be lighter than metal frames, and more durable to boot. Cyclists will spend less energy moving up hills, which will let them improve their times on race courses. For those of us not seeking to be Olympic-level athletes, reduced-weight bikes make commuting by bicycle easier, which is an important prerequisite in increasing the number of bicycle commuters in a city. All of these seem like good ideas.

We next stop to look at the gadget wall to look for a replacement fitness monitor to replace the one that broke just last week. It seems the latest model doesn't require a dedicated charger; it is instead powered by graphene-enhanced batteries charged by just moving around![6] In fact, just about all of the latest outdoor clothes are designed to generate power while we are in the sun, charging not only our fitness monitors but our cell phones and other small electronics as well. All of these innovations are made possible by the graphene-enhanced batteries, supercapacitors, and circuits that perform with nearly the efficiency of a superconductor. The first of these items were all black and dark gray because of the embedded graphene. Small lines were visible in the fibers where wires ran throughout. But, as manufacturing picked up and demand grew for a wider variety of colors and styles, designers got creative. Now, the lines are invisible, and you can hardly tell the difference between a regular shirt and these enhanced workout clothes. Other workout clothes feature not only the power-generating enhancement but also take advantage of graphene's incredible heat-conducting property. Sewn into the fibers of the garment are strips of graphene intended to move heat away from your core more efficiently than traditional cotton or nylon will allow. You'll keep cool in hot weather while out for a jog, able to feel even the slightest breeze. The advancements don't stop there, though. Winter coats and snow pants will take extra heat from your core and funnel it to your extremities to keep them warm. Gone are the days of sweating through your shirts while your fingers freeze.

We pass by the fishing poles on the way to check out and, lo and behold, even they are boasting about the graphene used in their construction. Rolling our eyes and beginning to wonder how we managed to make anything before the discovery of graphene, we note that the sign advertises that the pole will bend and withstand even more extreme angles if we use their special "proprietary" tackle line (which happens to be twenty-five times stronger than the leading brand, and, of course, is made using graphene). Some of these claims feel spurious, but with what other amazing products we've seen today that use graphene, we actually believe it. Maybe we will keep an eye out for videos of people bending their rods into figure eights—just for fun. With it in everything from tennis rackets, to tire rubber, to the very athletic clothing options, graphene seems to be everywhere.

In fact, we are reminded that our new car doesn't need oil changes. It's strange to think that we don't need to bring it into a shop for an oil change—ever—because of the new high-tech lubricant filled with graphene-covered nanodiamonds. Cartoons in the commercials show these little balls wrapped in a sheet of graphene and how they all help the parts spin and slide past one another. It's been rated for the life of the engine—the closed system makes maintenance so much easier on everyone's schedule. In fact, with the decreased friction and wear and tear on the engine, gas mileage for the car is better than ever. Some of the new energy recapture technology has made traveling even smoother. (In addition to simply recovering energy lost during braking, as is common in today's hybrid and electric cars, those in the future will likely recover energy from the heat in the exhaust pipe as well.) It makes the car we traded in just a couple of years ago feel so "last century."

While in the store waiting on our kids to finish their shopping, we reach out to absentmindedly spin a skateboard wheel and soon realize it's not stopping. It's silent. And it just keeps on spinning. The reason? Yes, graphene has been added as a lubricant in the sealed bearings of the wheels. Graphene *is* everywhere!

Snapping out of the mesmerizing moment, we realize how much has changed since the introduction of this seemingly simple molecule. It's been able to change the world, fitting into everything from high-tech electronics to innocuous everyday items. *How did we make anything in a pre-graphene world?*

Public and private research into graphene will continue to vigorously drive the next two decades of scientific advancement. With seemingly endless applications into which it could be inserted, the material promises to deliver a new world of abundance through efficiency and robustness. But there's that word again. Promises. It's all big talk until science actually delivers. All of these fun inventions sound like just nifty toys and conveniences now. But what if the impending revolution in medicine and water purification were brought to the developing nations? Imagine if all of these people had equal access to proper care and clean resources for building infrastructure without a messy nineteenth-century-style industrial revolution? Graphene isn't just about nifty gadgets and "gee whiz" parlor trick moments. To borrow from William McDonough and Michael Braungart—graphene will help us "Remake the way we make things."[7]

Chapter 6

GRAPHENE SUPERCHARGED

POWER TRANSMISSION

Graphene is not a traditional superconductor, but it is close. A low-temperature superconductor, as its name implies, conducts electricity without loss at low temperatures—very low temperatures. In 1911, Dutch physicist Heike Onnes discovered curious properties of some materials when they are cooled to temperatures approaching absolute zero (~4 Kelvin or –269°C): their electrical resistance drops to zero (not approximately zero, but truly zero, as in there is no resistance), and they repel, or eject, lines of magnetic flux (they keep the magnetic field from penetrating). The temperature at which these effects occur is said to be the material's critical temperature (T_c).

Why is this important? Because we waste a great deal of the electricity we produce in transporting it from the place at which it is generated to the user. The amount of loss depends upon the resistance of the metal, which, as its name implies, resists the flow of current through it. Metals tend to have lower resistances than other materials, which is why we use them in our electrical appliances. You experience these losses in everyday life when you notice the power cord of a space heater or hair dryer getting warm. These uses for electricity to generate heat are intentional conversions. The materials in hair dryers or space heaters are intended to get hot from the power coming from a socket. Other losses are less noticeable, or at least less attributed to waste. Has your phone ever gotten hot while you used it a lot? Resistance inherent to the materials that make up the phone cause it to heat up while under stress. Incandescent lights throw off lots of heat—they get up to several hundred degrees Celsius (still several hundred degrees Fahrenheit). Some of you reading this may remember having the Easy Bake Oven or Creepy Crawlers as kids. There is a reason why they worked so simply, and it was due to an incandescent light heating the cake or critter. The thermal energy, heat, is produced as the electrical current in the wire encounters the resistance of the wire in the device that is intended to produce heat. In a device that isn't built to heat something, unlike the aforementioned examples, heat is energy lost to resistance, an inefficiency in the system. The holy grail of electrical physicists would be a material that has zero resistance even up to 37°C (about 100°F). That way, we could transport electricity from where it would be created cheaply (in very rural areas) to where it is needed most (in the most urban areas).

Everywhere in the world today, there are hundreds of thousands of kilometers of electrical power lines stretching in every conceivable direction, each of which loses energy at every centimeter as it conducts electricity to our homes, offices, and manufacturing facilities. According to the US Energy Information Administration, transmission and distribution losses in the United States totaled between 6 percent and 7 percent of all electricity produced. That doesn't even count the inefficiencies and losses in the appliances that use the electricity on the consumer side.

This is why superconductors are so enticing. With a superconductor, the resistive losses in power transmission would go to zero. The problem with superconductors is that they are notoriously difficult to keep working. If they get too hot, their performance as a conductor doesn't just slowly get worse as the temperature rises, it abruptly stops superconducting and becomes a traditional lossy conductor when it reaches its critical temperature. There is no gradient. Materials are either superconducting, or they're not. Then there is that magnetic flux criteria mentioned in the first paragraph of this chapter. Even if the superconductor is kept colder than its critical temperature, if it is exposed to a strong magnetic field then its superconducting state can be abruptly lost. The strength of magnetic field that destroys the superconducting state is called the Critical Magnetic

Field. Unfortunately, when using electrical devices, one of the reasons electricity is so darned useful is that we use it to create, or in association with, external magnetic fields that can often be strong enough to crash the superconducting effect. Keeping meters, kilometers, or even thousands of kilometers of wire made from a superconductor below its critical temperature is currently impossible to accomplish. Niobium, a favorite traditional low-temperature superconductor, has a critical temperature of 4 Kelvin. On a typical winter day in Rhode Island, the daily high temperature is about 30 degrees Fahrenheit, or 272 Kelvin. To remain superconducting, a niobium wire would have to be kept colder than the average temperature on Pluto! Building our power transmission infrastructure from traditional superconductors is simply not practical.

In 1986, so-called *high-temperature superconductors* were discovered. They are called "high-temperature" because they maintain their superconducting, zero resistance, state all the way up to a balmy 90 to 130 Kelvin (–297 to –225 degrees Fahrenheit) or more. Breakthrough! Made from ceramic materials blending several unusual elements, high-temperature superconductors were all the rage as scientists and engineers raced to find ways to make and use large quantities, with the goal of infusing the technology into the energy infrastructure to realize the theoretical savings of a superconductor but without the high overhead of having to keep it super-cold. High-temperature superconductors could be kept cold using relatively common liquid nitrogen, which is much easier to produce and store than the liquid helium that is required for the traditional superconductor cousins. Liquid helium is several orders of magnitude more expensive than liquid nitrogen. Interestingly, at the right industrial volumes, liquid nitrogen becomes cheaper to buy per volume than distilled water. Unfortunately, widespread use of these high-temperature superconductors did not arise, largely because 90 Kelvin is still darn cold and difficult to maintain over large distances. This new class of materials also did not lend itself to the mass production of wires with the needed characteristics. Both types of superconductors are widely used in niche applications, but not on a massive scale, and certainly not (yet) in our energy transmission infrastructure or in everyday appliances.

Enter graphene. Graphene is not a normal-temperature superconductor. It doesn't have a critical temperature or a critical magnetic field strength sensitivity. Nor is its resistance to the flow of electrical current zero. But it is darn close. Close enough that engineers take notice and many are considering how its electrical properties can be used to reduce that 6 percent loss figure to something much lower. With an electrical resistance of less than silver, one of the most efficient electrical conductors, graphene is poised to become more widely used in all aspects of our power generation, transmission, and utilization infrastructure. And its electrical resistance doesn't vary all that much with temperature.

POWER STORAGE

Does is make you feel safe to know that you are likely carrying around containers of highly corrosive acid in your pocket or purse? How about within your car? Batteries. The mainstay of our modern, connected, and electrified world is batteries. They are also the Achilles' heel of the mobile power infrastructure. Ask any electrical engineer who has studied the power grid, and they will likely tell you that the one technology that hasn't seen much improvement is energy storage. We're still basically using the same chemistry-based approach to storing electrical power that we used fifty years ago, with only marginal improvement.

Batteries work by chemistry. To produce electrical power, they need to be able to store electrons and release them in a controlled manner as they are needed—not too much at any given time, nor too quickly. The negative terminal of the battery is the source of electrons that flow through the wires connecting your devices to it. When you begin to use, or draw, this current, negative ions flow through the liquid in the battery, depleting some of its stored energy. Fortunately, most batteries today are made from rechargeable materials so you can operate them in reverse: add electrons to the liquid to regenerate ions that are then are stored until needed. It is chemistry, and it works. It is also terribly inefficient, bulky, and the main reason your laptop computer weighs as much as it does.

Recharging batteries can be problematic. Recall the recent cell phone battery debacle in which phones melted, and sometimes exploded, for no apparent reason. It is important to remember that whenever you have a battery you have a potential bomb. The only difference is the rate at which the energy is released: slowly for a battery; quickly for a bomb. We don't like to think about that as we fill our pockets with the latest power-hungry compact electronic gear.

There are also fuel cells, which produce electricity through a different chemical reaction, but they suffer from many of the same problems: they are based on chemistry, heavy, and all-too-often dangerously explosive.

These types of batteries are great for small appliance applications, from your cell phone to your car, but they aren't very practical for large-scale use like for what would be required to store power produced during the day at a solar array farm so it can be available for customers to use after the sun sets and power generation stops. For storing a lot of energy, engineers have been more creative, but not creative enough to have a practical, universal solution to the long-term storage problem.

Consider molten-salt batteries. These batteries are large-scale and can be used to store thermal energy (heat) generated during the day by solar concentrators so that it can be used at night to generate electrical power. It is a neat idea and is being used as part of solar-thermal power generation sites around the world. But it suffers from the same drawback facing the large-scale solar-power generation industry in general: it is only practical in locations with plentiful sunlight and lots of underpopulated land. That rules out widespread use globally.

There are also gravity batteries. Hydroelectric power stations have a water reservoir located somewhere above the turbines. Late at night, when power consumption generally decreases as people go to bed, offices shut off their lights and adjust their thermostats to conserve electricity. Some hydroelectric dams will turn on pumps to carry water from the river upon which they are located to the reservoir above them. During the day, when electricity consumption is at its peak, and thereby at its highest price, they allow the water to flow downhill, pulled by gravity, to turn the turbines and generate more electrical power. To be clear, there is a separate reservoir, typically above the local water level, that is filled at night and drained during the day. This is in addition to the lake formed by the dam. (Talk about creative engineering to maximize profits!) This kind of battery is only practical because of the difference in local electricity prices between night and day, but it works.

So how does graphene play into this story? Graphene has some properties that make it an excellent candidate for use in something called a capacitor. A capacitor is a type of battery that isn't based on chemistry but on the idea that you can store energy in an electric field by separating two conducting plates with a nonconductor, called a dielectric. When charged, an electric field develops between the two plates, causing one to be positively charged and the other negatively charged. Because the dielectric isn't a conductor, current doesn't flow. The charge builds up, which means that the energy is stored until a critical threshold is reached. Eventually, any dielectric will break down and conduct electricity if the electric field strength gets too high. Different capacitors have different designs and different energy-storage limits. (Unfortunately, the bomb analogy holds for capacitors just as is does for chemical batteries.)

The energy-storage limit of a capacitor is proportional to the surface area of the conductive plates and inversely proportional to the distance between them. The larger the surface area of the conductor, and the more tightly packed together the plates, the more charge that can be stored. The ideal capacitor, often called a supercapacitor, has plates with large surface areas that are very close together. Now you can see why this discussion is in a book about graphene. Graphene is highly conductive (the right electrical property for making the capacitor plates), strong for its size (allowing the plates to be very thin and lightweight), and thin (allowing many plates to be stacked together in a small volume, increasing the available stored energy). Graphene may be the material that enables us to make true supercapacitors.

How much better might a graphene-enhanced capacitor be than a traditional battery? A lot better. Researchers at NASA are developing high-power-density capacitors called ultracapacitors that use folded graphene sheets to maximize the available surface area to store electrical charge in very small volume and at low mass. Figure 6-1 shows the results of a NASA study comparing conventional chemical batteries to ultracapacitors and graphene-based ultracapacitors.[1] Those made with graphene have energy densities comparable to chemical batteries but with more than a hundred times larger power densities. This means that graphene-enhanced batteries drive high-power systems for longer periods of time than any chemical battery. In addition, they can be rapidly recharged without the risks associated with rapidly charging chemical batteries. In other words, if you charge them quickly they won't melt or explode—failures that are all too common with today's high-power chemical batteries.

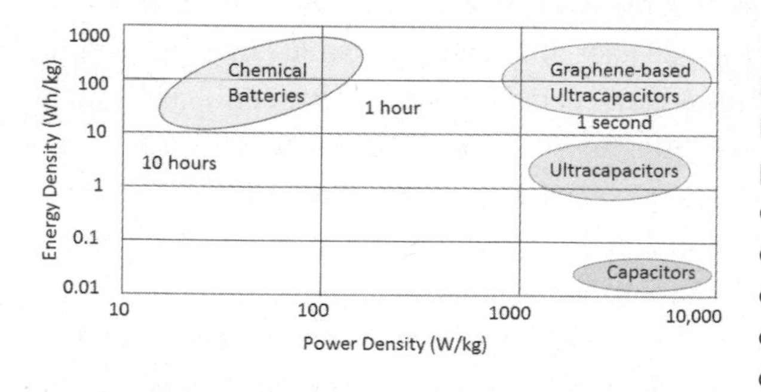

Figure 6-1: Graphene-based ultracapacitors have superior performance when compared to just about all other types of energy-storage devices. (Image courtesy of NASA.)

In practical terms, once the technology is perfected, batteries for consumer electronics will get much smaller, last longer, and be more easily rechargeable. Batteries on industrial scales will become more viable, allowing localized renewable power generation and storage to be practical for the first time. Homes might truly be able to generate and store enough electricity during the day, using solar power and graphene supercapacitors, to be removed from the grid. Several companies are investing in this technology, and the first products are already on the market.

POWER GENERATION

We will eventually have to wean our civilization from fossil fuels, and the sooner the better. From climate change, to the volatile politics surrounding many of the regions from which the world's oil flows, to the fact that we will eventually run out of readily accessible sources, the reasons for finding alternative energy sources are many. Unfortunately, for a variety of reasons, it won't be quick or easy to develop alternative sources that can meet our current and projected energy demands.

The most obvious source of alternative energy is the sun. All other sources of power, save for nuclear power, stem from the sun's energy in some way. Every square meter of the Earth receives approximately 1,361 watts of power per square meter whenever the sun is shining. If it could be perfectly collected and effectively harnessed, then the amount of energy falling on the Earth in a single hour of a single day could power the entirety of human civilization for a year. A single hour. But we don't, and cannot, have solar collectors operating at 100 percent efficiency covering every square meter of the planet to collect this energy. And, even if we could, then we would face the problem described above—how would the energy be stored so that it could be used when needed?

That doesn't mean we shouldn't implement solar power generation everywhere it makes sense. Some areas of the planet receive plentiful sunlight most days of the year and are excellent sites for building industrial-scale electrical power generation systems. Homes and businesses with solar arrays can take advantage of whatever sunlight they receive to offset their consumption of power from the grid, which most often is generated by fossil fuels. There is a lot to be done, but we're doing it very inefficiently. The state of the art for converting the energy contained in sunlight to useful electrical power is about 30 percent. That means that about 70 percent of the sun's energy that strikes a solar cell is not converted into power but is instead lost as heat or just reflected away. Surely, we can do better. And it looks like graphene may allow us to do just that.

When a particle of light, a photon, strikes a solar power generating cell, it knocks loose an electron, the charge carrier that makes electricity work. Not all photons create an electron and not all the created electrons are successfully transferred to become useful current produced by the cell. The laws of thermodynamics state that there are losses at every step, but anything we can do to minimize these losses increases the efficiency of the cell. These efficiency gains allow the cell to generate more useful power. Scientists in Switzerland have found a way to introduce selective impurities into graphene, in a process called "doping," that allows a single photon to produce up to two electrons instead of just one, effectively doubling the conversion efficiency of the cell to about 60 percent.[2]

But what about those pesky climates where it rains a lot? On cloudy days, there isn't enough sunlight to generate power using solar cells, making them useless and forcing consumers to find alternatives or go back on the grid. Right? Not necessarily . . .

Scientists in China had an epiphany. Recalling that graphene sheets can work very well as capacitors and as almost-superconductors, they thought about the fundamental physics involved in both and applied it to rainwater. Graphene's electrons are readily accessible (the reason graphene is such a good electrical conductor) so they readily attract positively charged ions. Opposites do attract! Given that rainwater is not pure water but contains all sorts of natural and manmade impurities like sodium, calcium and ammonia, many of these naturally ionized, or charged, it wasn't too much of a stretch to realize that they might be naturally attracted to the electrons in the graphene. If these oppositely charged ions could be separated into layers, then a natural capacitor would form every time it rained.[3]

The scientists tested their theory and created cells that produce electricity with an efficiency of about 6 percent. This isn't anything close to the efficiency with which solar cells convert sunlight to useful electricity, but it is far better than the alternative of creating no power on rainy days. It is also important to keep in mind that these are the first ever graphene power cells that use rain to generate electricity. The earliest silicon power cells that used sunlight to generate electricity were comparably low in efficiency, and it has taken decades to get them to the approximately 30 percent efficiency we see today. Following the theme of water/graphene power generation, another group of scientists in China noticed that when a drop of saltwater crosses a sheet of graphene it also generates electricity. Using flowing or falling water to generate electricity is not new.

Before people began harnessing electricity to run lights and machines, farmers were using the power of falling water to help them grind grains, miners were using it to drive pumps, and early industrialists were using it to grind just about anything that needed to be ground up. In the twentieth century, rivers and streams all over the world were dammed to build hydroelectric plants in the middle of the flowing water to generate electricity. The energy of the moving water is used to turn turbines, which, in turn, produce electricity. This means of power generation is carbon neutral, relatively inexpensive, and typically has minimal environmental impact. About 13 percent of the electricity produced in the United States comes from hydropower. What if you could produce useful amounts of electrical power on a much smaller scale? What if you could use rain water running off your roof to supplement your home's energy budget in a meaningful way?

Recall that salt (sodium chloride) easily ionizes in water, creating the positive charge carriers that can easily interact with graphene's accessible electrons. When the ionized saltwater flows across the graphene, it picks up some free electrons and redistributes them to the other side of the droplet as it flows, creating a voltage difference across the droplet. A voltage difference is what is required to make electricity flow, so this approach, on a very small scale, becomes a generator. If it can be scaled up, then the process might provide another method for individual, small-scale power generation, analogous to a hydroelectric dam but without the need for a huge river, massive turbines, and all the associated infrastructure.

Heat has long been used to generate electricity. In a nuclear-, coal-, or natural gas–fueled power plant, for example, steam is generated by the heat produced in the nuclear reaction or through the burning of the coal or natural gas. The steam is then used to turn turbines to generate the electricity we need. Each of the above methods are complex and require a sophisticated infrastructure to keep them running. Coal-fired power plants can require entire train loads of coal, daily, to keep running. Those using natural gas are typically connected to a major gas pipeline with the gas flowing continuously, twenty-four hours a day. And the operation of a nuclear plant is even more complicated due to the highly dangerous aspects of the nuclear fuel and the consequences should the power plant experience a major failure.

Scientists in Hong Kong have found a different way to use heat and graphene to generate electrical power, in a method that could be considered either a power-generation system or a battery.[4] Remember our discussion of graphene's loosely bound electrons responsible for its highly conductive properties? The scientists came up with a way to generate electrical power passively, by simply connecting a lower power light-emitting diode (LED) by wire to a piece of graphene immersed in a copper chloride (another type of salt) solution. The LED lit up; power was being generated by the graphene conductor in its interaction with the liquid.

The leading theory for how this works is that the copper chloride salt solution contains ions—unbound positive copper ions and negative chlorine ions. The copper ions are moving around in the liquid rapidly, due only to their ambient temperature. We're not talking about superheated liquids here; these are solutions kept at room temperature. As the copper ions bump into the graphene strip, they kick one of its loose electrons free. This free electron follows a rather simple rule of life, which also applies to electric circuits: always expend the least energy and take the shortest and easiest path to ground. In this case, the easiest path for the now-free electron is along the highly conductive graphene strip instead of out and through the copper chloride salt solution, which is also conductive. As it moves along the graphene sheet, it produces a voltage that in turn lights up the LED. We now have a rudimentary power generator or battery, depending upon your point of view, that is completely passive. The liquid will continue to absorb heat from the air around it, allowing continual replenishment of the liquid's thermal energy that causes the ions to move around in the first place.

SEMICONDUCTORS

A semiconductor is, as its name implies, a conductor that can conduct electricity under some conditions but not others. This is why they are useful. Semiconductors can rapidly be switched between conducting and not conducting states, allowing a binary on/off or 0/1 code, known as binary, to be used—which forms the basis of the information technology revolution we have experienced in the last sixty years. Semiconductors can be made to carry current in only one direction. They can be sensitive to light, pressure, heat, or other changes in their environment. Several different components connected together can respond differently under each circumstance.

Semiconductors are found in every aspect of our modern lives, from the obvious examples of the cell phones in our pockets and the computers on our desktops to the control systems that run our cars, refrigerators, and most home appliances; semiconductors are everywhere. It is difficult to imagine our modern world without the gadgets using semiconductors as a major part of it.

Graphene alone is not a semiconductor; it is a nearly super conductor. Something has to be done to make graphene function as an efficient semiconductor and that "something" is likely to be the addition of another element or chemical, in a process called doping. It is interesting to consider the irony in this. Most semiconductors used today are not, in their undoped state, conductors—they are insulators, or nonconductors. Consider silicon, the most famous element from which semiconductors are made, and after which the famous Silicon Valley near San Francisco is named. Unlike metals and graphene, silicon is a poor electrical conductor because it has no free electrons to conduct current; the outer shell electrons in silicon are tied up, bonded, so they cannot move around. To make a silicon-based crystal semiconducting, it must be doped with another element.[5]

Scientists love to give processes and conditions names. In chemistry, the name relevant to the discussion of doping is the "Octet Rule." According to the Octet Rule, an atom is stable when it has eight electrons in its outer shell. Think of an atom's shell as its skin. Each layer of skin can have only a certain number of electrons. If an atom has fewer electrons in its outer shell than are allowed for that layer, then it can readily share electrons with neighboring atoms to fill its shell. Once it does this, it is not likely to further react with other elements and is considered stable. This is the basis of chemistry.

Silicon has four electrons in its outer shell and readily shares electrons with other silicon atoms that surround it, forming a symmetrical-appearing lattice (figure 6-2). Each silicon atom is sharing spaces in its outermost shell with other silicon atoms, satisfying the Octet Rule, making them all content, and without unpaired electrons—causing silicon to be a nonconductor. To make it a semiconductor, scientists insert into the lattice either an atom that has five electrons in its outer shell or an atom that has three outer electrons. When an atom with five electrons is added, four of the five electrons bond with its neighboring silicon atoms, satisfying the Octet Rule, but this leaves one unpaired electron that is then free to move around. The free electron allows the new lattice to conduct electrical current, albeit poorly. This is called a negative or *n*-type semiconductor. Had the scientists doped it with an atom containing only three electrons instead of five, then only three of the neighboring silicon atoms would satisfy the Octet Rule and one would not. The unfulfilled silicon atom, the one that has no electron to fill its outer shell, then behaves like it is charge positive, attracting any free electrons roaming around to fill its shell. This type of semiconductor is called a positive, or *p*-type semiconductor. This is all standard stuff in the semiconductor manufacturing world, but doping a conductor like graphene to make it a semiconductor is not.

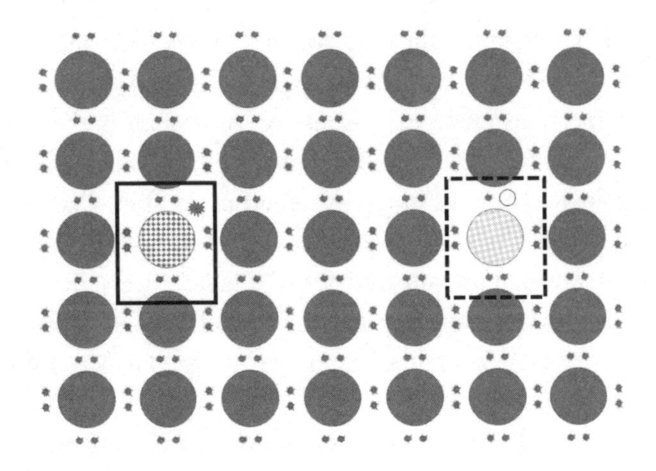

Figure 6-2: (Left, solid box) *N*-type dopant; note the extra electron forced into the lattice. (Right, dashed box) *P*-type dopant; note the hole where an electron is expected. (Image by Joseph Meany.)

With graphene, we need to find a way to a make a very good conductor into a nonconductor, at least when we want it to be. Following the silicon example, scientists are working to find a way to dope it, adding another atom or molecule into its lattice to affect its conductivity. This has successfully been accomplished with two-layer graphene, as opposed to the ideal single-atom, one-layer graphene so often discussed. In an approach pioneered in South Korea, bilayer graphene is doped on one side with an *n*-type dopant and simultaneously with a *p*-type on the surface of the other layer. The *n*-type dopant adds electrons to one side while the *p*-type attracts them on the other. This attraction results in the creation of an electric field, as is the case whenever positive and negative ions or materials are brought close together, and this field induces a bandgap, or a region in which no electron can exist—and therefore no electrons can flow, making the material, effectively, a nonconductor.[6] Voila. We now have a semiconducting dual layer of graphene.

It appears that graphene can also be made semiconducting in ways that don't require doping. Recall that the shape of graphene, being planar and essentially only two dimensional, plays a key role in making it such a good conductor. Graphene's two dimensionality also makes it vulnerable to having the flow of current interrupted if the flatness is somehow disturbed or wrinkled. On a small piece of graphene, of the size that might be of interest to the electronics industry, a wrinkle can cause the graphene's conductivity to be interrupted (nonconducting) in one direction, making it a semiconductor.[7]

Finally, researchers at Georgia Tech found that by folding graphene sheets multiple times into ribbons, they can turn it into a semiconductor. Looking like perfectly spaced waves on the surface of the ocean, graphene waves only 400 nanometers apart, resulting from precision folding, function as semiconductors for reasons not yet clearly understood. The method they use to "grow" the waves, the semiconducting portion of the graphene, allows for many such regions to be formed on a graphene wafer—making the transition of the technology to the creation of working transistors potentially much easier.[8]

In this chapter, we've looked at how we might use graphene's inherent and fantastic electrical conductivity to improve the efficiency of power generation, transmission, storage, and utilization. We've also looked at how researchers are working toward modifying its structure so that it can also function as a semiconductor, taking graphene a step closer to enabling its use in the next generation of electronics.

Part Three

WINNERS AND LOSERS

Chapter 7

DISRUPTION

Technological disruption is not new. Each technological advancement leading to our modern lifestyle is a direct result of disruptions in years past—some long past. The invention of agriculture allowed the development of civilization—as opposed to simply individuals and groups subsisting in hunter/gatherer clans across the African landscape. Our modern mass agriculture is merely an efficient extension of the original innovation that allowed some members of society to plan their meals ahead of time rather than merely searching for a meal each day. If you can rely on someone else to provide the food that you are going to eat in any given day, then you can have the luxury of time to contemplate broader problems, such as how to improve sanitation (another disruptive and beneficial innovation), medical care, and eventually computers and rocket ships.

In this chapter, we will look to historical examples of disruptive innovations, how they were initially received, and their overall impact on society and the world that evolved after their introduction. We will then examine the disruptive qualities of graphene and attempt to see how it will do the same, hopefully making for a better tomorrow in the process. We will begin with an invention originally heralded as "a solution waiting for a problem," the laser.

LIGHT AMPLIFICATION BY STIMULATED EMISSION OF RADIATION (LASER)

In 1958, two scientists at Bell Labs filed a patent application for what they called an "optical maser," and which later became known as a LASER, or, as is now common today, simply a laser.[1] When Charles Townes and

Arthur Schawlow filed their patent application, they believed their invention had many potential practical applications, but virtually none of them were immediately realizable. At about the same time, Gordon Gould, a graduate student at Columbia University, proposed essentially the same type of device and postulated its use in interferometry, radar, and nuclear fusion. These were all good ideas, but none were mature enough to see the laser applied to them any time soon.

Fast-forward sixty years and it is difficult to imagine our world without lasers. They are used in space exploration for determining the distance to the moon and other space objects, at the grocery store to read the bar codes for pricing and inventory management, by eye doctors for correcting our vision, for transmitting huge amounts of data via fiber optic cables, in police speed detectors, in our CD and DVD players, and at home to provide endless entertainment for our cats. This is just the tip of the iceberg, and it is important to note that almost none of these applications were or could have been foreseen by the inventors of the laser when they filed their patent application. The laser was a neat invention waiting on practical application. *It was disruptive.*

THE MICROPROCESSOR

The Electronic Numerical Integrator and Computer (ENIAC) may not have been the world's first electronic computer, but it is the device that most people consider as such.[2] Commissioned at the University of Pennsylvania in 1946, ENIAC was used primarily by the US military to compute more accurate artillery firing tables for the US Army. But ENIAC did more than the jobs it was built to accomplish. It inspired an entire generation of visionaries to pursue the development of computers, leading to the mainframe computer revolution of the 1960s and 1970s, the microcomputer and personal computer revolution of the 1980s and 1990s, and the smartphone revolution of the 2010s. Today, we see computers being integrated into almost everything around us; from our appliances and automobiles to our home heating and cooling systems and the very clothes we will be wearing, the microprocessor is finding its way into almost everything. *It has been and will continue to be disruptive.*

THE INTERNET

If you remember the visceral pleasure of drinking a cup of coffee while reading a good old-fashioned newspaper, then you know one of the things we've lost with the implementation of the internet. If you run a brick-and-mortar retail shop, where you sell just about any "thing," then you know all too well the impact of e-commerce via the internet—and you've learned to use it, or you most likely wouldn't still be in business. Just ask the managers of the thousands of bookstores around the world that have closed due to the invention of online physical bookselling and e-books. Do you remember record stores? Now we have iTunes. The list goes on and on. Modern communication and retailing is vastly different than it was just a couple of decades ago, and businesses were forced to adapt quickly or die. Indeed, many died. Even industries not traditionally associated with retail have had to adapt to save money and remain competitive. For example, many companies outsource their copyediting and reviewing to less expensive employees from the Philippines, Thailand, or India. The internet changed how we get our news, how we select our mates (dating apps), our political and social discourse (Twitter, Facebook), the way we plan our travels (Travelocity, Kayak, etc.), and the very way we spend our time both at work and at home. One of the most popular home electronics items is Amazon's Echo. Echo is a home assistant, voice activated, that can control virtually all aspects of your home with a simple voice command, thanks to its super-fast microprocessor (disruptive technology) that is linked to the internet (disruptive technology). Just say, "Alexa, what is the news?" and you get your daily news briefing from the news outlet of your choice. You can ask Alexa to add items to the grocery list, to order you a pizza (charged, of course, to your credit card on file), to dim the lights, turn up the air conditioner, and lock all the doors. *The internet was and continues to be disruptive.*

The list of modern disruptive technologies is long: digital data, made possible by the internet, is disrupting the entertainment industry (MP3 music, video on demand, e-books, etc.). Fracking is disrupting the global production and distribution of oil, changing longstanding political power structures in the process. Solar photovoltaics are allowing affordable and decentralized power generation, disrupting the established power generation and distribution infrastructure. Satellites changed the way we communicate with each other (satellites as relay stations), the way we navigate (the Global Positioning System), the way we predict the weather, and how we fight wars.

What these disruptive innovations have in common is their successes. But what about the innovations that were thought or believed to be disruptive but weren't?

DO YOU REMEMBER HIGH-TEMPERATURE SUPERCONDUCTORS?

Wires, made of metal, are what we use in our everyday lives to conduct electricity to the many devices around us that require it. Metal wire carries the power from the power plant where it is generated, to the utility poles that crisscross our neighborhoods, to the wires in the walls of our homes and, to the outlets into which we plug our appliances. And at each step there are losses. At the most basic level, these losses are caused by a property of the wire conductor called its resistivity. Aptly named, resistivity is a measure of the how much the wire resists the flow of current, turning the inefficiency of its transmission into waste heat. Because of its flexibility, relative mechanical strength, and relatively low resistivity, copper is the metal of choice for most of our electrical power transmission systems.

In 1911, Dutch physicist Heike Kamerlingh Onnes discovered that some materials, when cooled to extremely low temperatures, conducted electricity without any resistivity—meaning there would be no loss in current over whatever distance the current was conducted.[3] These materials, called superconductors, are discussed in chapter 6. Unfortunately, their practicality was severely limited by the low temperatures, approximately 4 Kelvin, or −452 degrees Fahrenheit, at which they operate. This was a problem and limited superconductors to highly specific niche applications until 1986, when Georg Bednorz and K. Alex Müller discovered a new class of superconducting materials that had only to be kept at 128 K (−211°F), a significant improvement and one that could be practically implemented on a massive scale using state-of-the-art industrial coolers.[4] The scientific journals and popular press were flooded with research papers describing the many ways these high-temperature superconductors would "change everything."[5] They were going to change the world be part of every element of our power infrastructure.

Alas, this was not to be. These new superconductors were ceramics, not metals, and therefore did not have the mechanical properties (see chapter 1) we so desire in our electrical power systems. Making wires from them is impractical; keeping wires cold across tens, hundreds, or even thousands of miles is still a significant challenge. And, like their lower temperature cousins, high-temperature superconductors lose their miracle conductivity if the current they carry gets too high, if they get too warm, or the local magnetic fields get too large. Soon, the predictions of the high-temperature superconductor revolution began to fade. They are still promising, but few people, if any, are now predicting they will change our world.

AND THEN THERE WAS COLD FUSION

Fusion is the holy grail of nuclear engineers trying to come up with green, plentiful power to replace our reliance on fossil fuels. There is nothing magical about fusion. The sun uses its tremendous mass, and the pressure at the center of the star that this mass generates, to squeeze hydrogen atoms together until they finally merge and become helium, releasing energy in the process. Therefore, the sun shines, releases energy, and doesn't collapse under its own weight. We currently use nuclear fission, the splitting of atoms, in our electrical power plants. While they are carbon neutral, nuclear power plants aren't exactly clean—each power plant produces dangerous and toxic nuclear waste that must be managed and safely stored for hundreds or thousands of years. We can also use fission bombs, like those dropped on Hiroshima and Nagasaki in World War II, to initiate the fusion process and make bigger bombs, generically called hydrogen bombs. In more peaceful scenarios, scientists use high-power lasers to start the fusion process in facilities like the National Ignition Facility and at ITER (which means "the way" in Latin), an international fusion research effort located in France. The problem with fusion, as with superconductors, is that making it practical has turned out to be very difficult.

That's why in 1989, when Martin Fleischmann and Stanley Pons reported that they had measured "excess heat" that seemed to indicate nuclear fusion was occurring, using simple bench chemistry rather than extremely high energy physics, many could envision the clean energy future offered by nuclear fusion coming to a reality much more quickly than anyone had anticipated. Their experiment was simple: they ran an electrical current through a specially prepared type of water called "heavy water" on the surface of a palladium electrode and, voila, excess heat that could not be attributed to simple chemistry was present. Unfortunately, their results could not be readily duplicated, and it soon became clear that they had made their claim prematurely.[6] Upon closer examination, it was clear that they had not appropriately accounted for sources of error in their experiment, nor had they detected nuclear byproducts that would have to be there had fusion occurred. "Cold fusion" was the term used to describe this process, since it didn't require tremendous energy input to make it happen. But cold fusion, with all its promises, was effectively dead.

LEO BAEKELAND AND HIS DISCOVERY OF PLASTIC

Finally, let's talk about a *successful* disruptive technology that seems to be the closest parallel to graphene in the modern era: plastic.

"There's a great future in plastics . . ." said Mr. McGuire to Ben, the character played by a young Dustin Hoffman in the movie *The Graduate*.[7] And Mr. McGuire was correct. For better or worse, plastic has changed our world.[8] Like graphene, plastic is made from carbon-based molecules. Long chains of carbon atoms and other elements linked together in a repetitive sequence are generally referred to as *polymers*, which is just the fancy word for plastic. Plastic bottles and shopping bags are not the only polymers, though. There are natural polymers, like starch, proteins, or DNA, which make your body function. Polymers were discovered a long time ago, but they weren't put into significant use until after they were first made from fossil fuels—to be specific, oil—just after the beginning of the twentieth century. The first of these plastics was invented by the curious character Leo Baekeland, and is now known as "Bakelite." From this first modern, synthetic plastic came a familiar litany of others: polystyrene, polyester, polyvinylchloride (PVC), polythene, nylon, and polyethylene terephthalate (PET), to name a few.

Leo Baekeland's story is a very typically American one—an immigrant comes to America and becomes rich. Baekeland was born in Belgium in the middle of the nineteenth century and moved to New York in 1889 to study chemistry. After he completed his studies, he decided to remain. He was an inventor, and he became wealthy after one of his photographic inventions, a type of photographic film, was sold to Eastman Kodak for the amazing sum of $750,000. This is a lot of money today, so imagine what it was worth in 1898! Baekeland then turned his creative thinking toward the problem of creating a synthetic form of shellac. At that time, the only way to make shellac was to take the resin secreted by the female lac bug and dissolve it in ethanol. This was time consuming and expensive. Surely there had to be a better way. In this age of innovation, Baekeland was on the case.

After several false starts, failures, and other missteps in his quest to produce artificial shellac, Baekeland inadvertently made a polymer that he then tweaked to become a hard moldable material, plastic, which he called Bakelite. Soon, Bakelite was being used everywhere. Like he had his photographic invention, Baekeland sold his Bakelite business as well, this time to the Union Carbide and Carbon Corporation, which we now know as simply Union Carbide. It was the Union Carbide company that later employed Roger Bacon, our famous inventor of the carbon fiber in 1959 (discussed in chapter 2).

It is worth remembering that timing is everything. Baekeland was not the only inventor working on the shellac problem, nor was he the only one combining various organic chemicals together in search of new compounds and resins. British inventor James Swinburne was working on a similar problem, also discovered plastic, and lost the patenting race to Baekeland by a single day!

Plastic is used in nearly everything, just as we imagine graphene will be. We drink our water from bottles made from PET (Polyethylene terephthalate). We wear clothes made from nylon and polyester, drive cars with plastic parts throughout, fly in airplanes lined with plastic overhead bins, and use radios, televisions, and computers encased in streamlined plastic cases. We carry our groceries in the ubiquitous plastic bags (that also happen to be a blight upon our landscape—so much so that some states charge a tax every time you get a new one from a grocery store). Our pens are made from plastic. There are plastic parts in most, if not all, household appliances. Plastic gears replaced metal ones in our windshield wipers, household mixers, and handheld power drills. And, of course, we all sit in those uncomfortable white plastic lawn chairs throughout the summer months.

Here's an interesting statistic from the European Association of Plastics Manufacturers: in 2014, European building and construction endeavors used more than 9.6 million tons of plastics.[9] Plastic was used as insulation, pipes, and window frames, in addition to the smoke detectors, smoke alarms, electrical outlet covers, light fixture housing, etc. According to the

Worldwatch Institute, 299 million tons of plastic products were manufactured in 2013, generating over $600 billion in revenue, with the average person in North America and Europe consuming nearly 100 kilograms of plastic per year. The use of plastic products is increasing rapidly in China and India, so worldwide use is expected to continue to grow.[10] Just how much plastic is that? About 100 billion pounds per year! In 2013, 107.5 billion pounds of plastics and resins were manufactured—an increase over the previous year's 105.9 billion pounds. You get the picture. Plastic is in just about everything. *As graphene will be.*

So this takes us back to graphene. Will it truly change the world in the same way as plastic, the laser, the microprocessor, and satellites? Or will it go the way of cold fusion and high-temperature superconductors? Time will surely tell, but, if you believe the headlines and the many scientific research papers being published globally, the answer looks like it will be on the side of lasers and the internet. And there is another reason to suppose that this will be the case. Graphene seems to be a material with applications for just about everything we as humans do: electronics, building materials, optics, recreational activities and equipment, transportation, energy, and even space exploration.

What would you make if you had an extremely lightweight, flexible, break-resistant, low-friction material with a long life span?

Let's tackle the last part of the question first. Why do materials wear out over time? Who hasn't been frustrated when your favorite pair of jeans starts showing the inevitable thinness that leads to them developing holes? Or when your kitchen blender finally stops working because the gears that allow the motor to vary the speed and strength of the blending wear down and break? The life limiter on more than one of the cars I've personally owned was the transmission. Over time, friction takes its toll, and the gearing simply breaks.

To understand how graphene might be able to alleviate this problem, we first need to better understand how and why friction occurs. When surfaces rub against each other, the actual contact points are only nanometers in size—just a very few atoms. Friction is greatest when the stiffness of any surface protrusions is roughly average. In other words, when it isn't too soft or too stiff—both extremes can decrease the relative friction. Determining the actual underlying cause of friction is quite complicated—you must consider the surface roughness, small variations in the shape of the material, and surface contamination. The study of friction, called *tribology*, is an extremely specialized area of material science.

When friction occurs, the energy of the moving surface is converted to thermal energy (heat), which can have some interesting or potentially damaging results. Aren't Boy Scouts and Girl Scouts supposed to be able to start fires by rubbing two sticks together? Tom Hanks, in the movie *Castaway*, learned how to start a fire by doing just that.[11] In a more modern setting, the moving parts in your automobile engine and the heat they generate from friction as you drive, is the primary reason you should use motor oil and have a cooling system. Otherwise, the heat generated by the engine would quickly destroy the engine and perhaps cause the car to catch on fire. Over time, despite the use of the best lubricants, the friction within the engine causes material damage and the engine needs to be replaced. I could go on, but you get the idea.

This is where graphene might play an important role in reducing the friction of just about everything. We now know that graphene is super-strong, but only if the single-atom thick material contains no imperfections. This means that there can be no inherent surface roughness problems and the material is resistant to surface contamination. If it can be manufactured to precise shapes, the primary causes of friction may be removed. Graphene coatings have already been applied to small machine parts, allowing them to dramatically increase their operational life spans and produce almost no friction-related waste heat. And there's more. In micromachinery, it might be possible to maintain atomic-scale alignments by selectively introducing contaminants in the graphene coating so that the preferred direction of motion has virtually no friction while movement in other directions does have friction. This passive self-alignment scheme is already being tested in the laboratory.

Of course, there are some wear problems, such as when you want to reduce the wear but not really decrease the friction all that much. A good example of this your automobile's tires. Tires come with ratings in terms of miles—about how many miles can you expect to drive, on average, before the tire wears out and needs to be replaced. For modern tires, this range is 40,000 to 90,000 miles. Generally speaking, tires with higher mileage ratings tend to be stiffer and tougher than tires with low mileage ratings, which are typically soft and provide very good traction with the road—traction begins the positive side of friction. We may not want to make a complete automobile tire from graphene, and it certainly doesn't make sense to just coat the outer layer with graphene, lest the tires have dangerously low traction. Who wants to drive on a surface with the friction characteristics of ice? For these reasons, manufacturers are incorporating graphene flakes in their tires to provide added wear resistance and strength without compromising performance. Designing long-life, high-performance tires is a great deal more complicated than just adjusting the stiffness and robustness of the materials. Tire size, width, tread shape and depth, and inflation pressure are all major factors. Having the ability to tune these other attributes with a lighter-weight, stronger, and variable-friction material like graphene gives designers another tool for their toolbox.

Along with low friction, building objects and devices that are resistant to breaking is another revolutionary application of graphene. Given graphene's inherent strength (described in chapter 5, the possible applications are endless. When is the last time you broke or chipped your favorite ceramic coffee mug? What about those annoying rocks thrown up by a passing tractor trailer that dinged the paint on your new car? Remember that time you dropped your smartphone and cracked the screen? And there's always that plastic plumbing fixture you were trying to replace and ended up overtightening, either stripping or breaking it. Never again. Paints impregnated with graphene, or materials that have layers of graphene deposited on them, will have break resistances unimaginable today.

One of the ways products are made stronger and break resistant today is through adding mass—making the plastic or wood thicker so that it won't crack as easily, increasing the density to strengthen the material, or adding spars or extra fasteners to keep the material from being as stressed during use. These have one side effect in common—they increase the weight of whatever is being strengthened. That is a significant problem. People would love to have a cell phone that was unbreakable but are they willing to carry around a brick in their pocket to make that one feature a reality? Cars can be generally safer by using denser materials in certain places. But once you add mass, down goes the fuel economy. Using graphene instead of traditional strengthening methods can make items much more robust, while keeping their weight lighter.

Automobile engines and tires that for all practical purposes don't wear out, shoes that you keep until you feel like getting rid of them, machines that don't require frequent servicing from normal wear and tear, clothes that last a lifetime, carpets that don't start to show bare spots in front of doors and other high-traffic areas—graphene can improve the durability of just about anything.

Next, let's examine the "extremely lightweight and flexible" characteristics of graphene. It is here that many of the most fantastic promises are being made with regard to graphene's disruptive potential. Being made of single atoms arrayed in a flat plane, graphene is very thin and very strong for its dimensions. This means it can be bent, rolled, folded, and otherwise formed into just about any shape you can imagine. It can be stretched to about 120 percent of its original size without breaking, and it can snap back to its beginning state with ease. Add to this the fact that graphene transmits 98 percent of the visible light that strikes it, and you have a lightweight, flexible, electrically conducting, and nearly invisible material. Wow. Many applications, in particular those that require computer-like functionality, will certainly be thicker than one atomic layer, but not by much. They will still be nearly as flexible and transparent. What can you do with such a material?

For one thing, we may quickly move beyond today's smart phones and smart watches to integrated devices that we wear on our wrist, putting them on with the ease of those pesky snap bands that are so popular at carnivals and in toy stores. On your table or desk, you can use this smartphone-sized computer as you would any tablet or phone, to check your email, respond to the latest Twitter or Facebook posting, and catch up on the latest sports scores. When it is time to go, you can pick it up and snap it on your wrist, where it conforms and resides, ready for continued use.

Following this line of thinking, why constrain applications to the very small? Wouldn't you like to have your living room wall completely covered by a transparent, graphene-enabled television or computer screen that waits, invisible, until it is activated? While we wait on true *Star Trek*-like holodeck technology to be perfected, we can ditch the virtual reality glasses and use rooms that have every surface covered in graphene image projectors to place us, visually, anywhere we want to go: flying over the Grand Canyon or through deep space, walking through the Vatican, or even taking a stroll across a field mapped anywhere in the world, as services such as Google and Apple Maps continue to photograph, in high definition, just about every part of the globe. Imagine the applications for employee training, crime-scene investigations, and the tourism business.

Don't forget that lightweight and flexible also means mobility. Graphene-enabled, thin-film computers, like those described above, could invisibly cover the window of your soon-to-be-self-driving car, providing maps and real-time traffic reports and routing as you navigate through New York, Los Angeles, or any city in between.

These flexible screens might be embedded into the very clothes we wear, allowing us to instantly change the color of our shirts from blue to red or form unique color patterns as we show our individuality at Saturday night's social event. As you're walking on a cloudy day and the sun comes out, you can shift your shirt color from dark to white to avoid overheating. You could even turn yourself into a walking billboard to advertise your business as you head down a busy street on your way to lunch.

If we're thinking of computer-like applications enabled by thin graphene sheets, why not think very small and embed computers into our contact lenses so we can have heads-up display technology and access to information privately streamed to our eyes anytime we wish? This could take the fine art of daydreaming during boring business meetings to a whole new level . . .

All of this brings to mind two additional properties of graphene that are both very useful and, in their application, highly disruptive: high electrical conductivity and thermal stability. Graphene, thanks to it being a single layer thick and made of carbon, can conduct electricity with much lower resistance than copper. This will make any electrically powered device much more efficient. Since less electricity will be turned into heat, which is wasteful, the power is easily conducted from one part of a device to another. This increased efficiency translates into what is the holy grail of the commercial electronics industry—longer battery life. But why stop with increasing the useful life of heavy chemistry-based batteries? It turns out that graphene can also be used to increase the efficiency and performance of another power storage device, the capacitor.

The electrical properties and promise of graphene are extensive. So much so that we devoted chapter 6 to them.

Chapter 8
OBSTACLES

You may have the best material, idea, or technology since the discovery of fire, but until you convince potential users or customers that your widget is better (for whatever reason) than the widget they are currently using, then it will not be readily adopted. Getting products made from graphene into our hands will not be easy. Aside from the usual stumbling blocks associated with manufacturing, marketing, and distributing a new or reformulated widget, graphene-based products have the added problems of creating and maintaining a supply chain of raw material, competing with technologies that have entrenched customer bases, and dealing with the inevitable lawyers. Being among the first to tread this brave new world is not for the faint of heart!

SUPPLY

Consider electric cars. The developed world has access to the electricity that will be needed to recharge the batteries of fuel-efficient electric cars. The batteries, though currently big and bulky, exist to make electric cars viable. A significant drawback to electric vehicles is their limited range of driving due to the power-storage limits. Any road trips using electric cars will require one of two things: 1) convenient, affordable, and geographically widespread places to stop and recharge the car's batteries quickly, or 2) affordable and geographically widespread places to quickly swap out depleted batteries for fully charged ones as they are needed—as quickly and easily as stopping to refill a gasoline powered car is today. Neither of these preconditions exist yet and, therefore, electric cars are rare and mostly only driven locally. Taking one on a road trip across the country is just not practical.

The promise of graphene is in an analogous situation. Tens of thousands of patent applications have been filed in a thousand different areas, which may lead to tens of thousands of innovative new products. Currently, making graphene is difficult (see chapter 4) and before it will be readily adopted it will have to follow through on promises to provide benefits far greater than existing technology or for a far lower price. It will also have to be available in the quantities customers want, when they need it, in sufficient quantity, and of reliable quality to be useful. We are like the driver of an all-electric car finding ourselves needing to drive from New York to Seattle but unable to make it due to the lack of electric car service stations along the way.

What's the status of graphene production? With companies around the world making graphene and new methods of producing it being discovered at an astonishing pace, it seems plausible that someone will discover a way to mass produce it at an industrial scale within a few years. Some will produce graphene in small, discrete quantities (think millimeters to centimeters in length, or less) for use as an additive or in conjunction with other materials. To be truly useful, and probably profitable, such production will need to exceed a few thousand tons per year. Others will produce single or few-layer graphene sheets from either raw ore or some combination of CVD and epitaxy. In this case, there is no standard or optimal size or area. Production will be driven by the customers' needs. So, companies will likely need to make it in variable areas, in quantities up to a million square meters per year or more.

Let us not forget cost. Having people make graphene the way it was originally discovered would be so labor intense (hence expensive) that the material would never become more than an intellectual curiosity. If graphene follows the trend of most other industrial materials, the first production runs will be expensive and support only niche applications. Think of the story about the first aluminum products. We are in the "coat royal utensils with graphene" phase of products, in which they can be sold at a premium price for either their superior performance or novelty. This is basic supply and demand. If a demand exists and the supply is low, then the cost will be high. With high materials cost, the graphene-enhanced product, no matter what it is, will have to sell for a higher price in order to allow the producer to recover their materials and labor costs. As commercial production increases, and as more producers enter the marketplace, the benefits of competition will start to manifest, driving down the cost of the material for the end user.

Let's look at fossil fuels. Regardless of where you stand on the environmental concerns of hydraulic fracturing—or fracking—the fact remains that the development of the process dramatically changed the fossil fuel industry and made the United States again among one of the top producers of fossil fuels in the world. Recall that fracking is a technique that allows otherwise inaccessible fuels like petroleum and natural gas to flow more freely after the rocks that contain or surround it are fractured by injecting a pressurized liquid. Fracking is expensive, and only made good economic sense when the price of fossil fuels was high enough to justify the cost of extracting, processing, and delivering fuel products made accessible by it. This was the case just a few years ago, when the price per barrel of oil reached the $100 mark.[1] US production of fossil fuels rose and rose until world fuel supplies began to exceed demand, causing the price to fall precipitously. This was not a shock to those of us who studied economics. Suddenly, the fossil fuels produced by many of the fracking wells cost more to extract than it could be sold for. New production ceased, people were laid off, and production leveled off, awaiting the next surge in demand that would make fracking profitable again.

How does this relate to graphene? Currently, the forecasted demand for graphene is high. With tens of thousands of new graphene application patents being filed per year and global production barely able to keep up with the demands of laboratory researchers, let alone the commercial marketplace, the price for high-quality graphene is relatively high. If the "killer app" for graphene production is found—meaning a commercial product that will be in high demand and therefore profitable—there will be a race to see who can produce graphene in sufficient quantities to meet that demand. Once the production ramps up, particularly if there are many suppliers, the price per unit (per gram or square meter) will drop, and the laws of economics will undoubtedly step in and establish a robust commercial marketplace for it.

The development of the Haber-Bosch process is an excellent historical example of the way this could play out. Modern mass agriculture would be impossible without an inexpensive and plentiful supply of nitrogen. Nitrogen is what makes plants healthy as they grow and is the primary ingredient, in one form or another, of fertilizers. This has been known for over 150 years and was the impetus for the industrialized nations of Europe to seek an artificial source of nitrogen so that crop yields could increase to feed their growing populations. (There are only so many cow patties to go around . . .)

In 1898, William Crooks, president of the British Association for the Advancement of Science, challenged the scientists of Europe to develop an industrial process to make nitrogen for fertilizer so that it could be mass produced and used in agriculture.[2] What followed was a tale of industrial secrets, a world war, and, of course, a Nobel Prize.

In 1909, just under ten years after Crooks's challenge was issued, a German scientist named Fritz Haber found a way to combine nitrogen and hydrogen into ammonia using high pressure and extreme temperatures. Fellow German scientist Carl Bosch then figured out how to mass produce ammonia using the newly discovered chemical process. Scientists at the time already knew how to convert ammonia into fertilizer, thanks to the work of Wilhelm Ostwald, so Haber's process was the missing piece of the industrial fertilizer production puzzle.[3] But then a little problem called World War I began, pitting Germany against England in a bloody struggle that raged across Europe. And, as is all-too-typical of modern societies, the industrial process that produced fertilizer was converted to produce a close cousin of fertilizer—explosives. The Haber-Bosch Process was now not only an industrial/commercial secret, it was a military secret as well.

After Germany lost the war, the secret Haber-Bosch Process was revealed and adopted around the world. Only a few years later, in 1920, Haber was awarded the Nobel Prize for discovering the chemical process that produced the ammonia.[4] In 1932, Carl Bosch and Frederick Bergius were awarded a Nobel Prize for the high-pressure techniques used in what was now known as the Haber-Bosch Process.[5] Today, more than two million tons of ammonia are produced each week globally, with the bulk of it being used to produce fertilizer. Chances are, the last meal you consumed before reading this chapter was made possible by fertilizer made from Haber-Bosch produced ammonia.

INERTIA

Imagine you are the manufacturer of a commercial product like tennis shoes, which would benefit from more durable, long-lasting materials. You have an existing production, sales, distribution, and financial plan that has milestones you must meet to remain profitable and solvent. These milestones might be the ones you expect: sales volume, gross revenues, stock dividends, or quarterly share price. They all depend upon keeping the demand up, the supply adequate, and the cost per unit profitable and affordable. Your latest tennis shoe design incorporates graphene to make it more durable and, because the graphene is blended in a composite material and not pristine, perhaps give those who wear it better traction than any other shoe on the market. To have the shoes in stores by next Christmas, you need to start producing them ten to twelve months in advance and put your marketing campaign (print, electronic, radio, and in-store ads) into place now. Will you take this leap if you don't already have a contract in place with a proven supplier who can meet your production schedule and quality demands? How reliable is this provider? Have they produced a similar quantity of a similar material for any other customers with whom you can check to determine if they met their obligations in the past? You can see the tennis shoe maker's dilemma. Will the market support their planned product?

A product can only be disruptive, enabling, or even useful if someone wants to use it. This may seem like an obvious statement, but in the business world the nuts, bolts, and final cost/benefit trade will either support the use of graphene or dismiss it. For many of the graphene applications and products being hyped, the verdict is still out.

With the uncertainties of being able to mass produce graphene affordably yet to be resolved, how will a manufacturer of a profitable product justify abandoning their time-proven supply chain and manufacturing processes to use state-of-the-art materials for an unknown marginal gain? Will companies keep going with what they have on hand, the lowest risk approach, until the supply of affordable graphene is in place? To the ears of a scientist, the tendency of a company to keep moving in the direction it is currently going unless acted upon by an outside force (like the disruption potential of graphene) sounds a lot like Newton's First Law of Motion. After all, the company has an existing customer base, and existing production cost model, workers trained in the current methods of production, and suppliers accustomed to meeting the company's needs. Is the promise of graphene enough to disrupt all this? These are questions that graphene suppliers will need to have answers for as they seek to carve out a niche.

Having worked at NASA for most of my career, I (author Johnson) have experienced this corporate inertia first hand. Sending a robotic spacecraft to any destination is difficult and expensive. Even with the cost of launch coming down, it is simply impossible to mount a space mission unless you have millions of dollars at your disposal. With that in mind, you know that your customer (the person, government, or corporation) who is funding the mission wants it to have a high probability of being successful. No one is going to spend millions of dollars and not care if the result is a failure.

To assure that success, the team designing the hardware for the mission will look at the requirements to determine what they need to design and select parts that will allow it to be built. The least expensive and least risky approach is always to select space-qualified hardware that has flown successfully in the past. Even if the mission is going to fly a telescope or sensor that is brand new and never before used, the support equipment must be as reliable as possible. To make my point, let's assume we're going to fly a new type of telescope to the moon.

Our primary goal, then, is to make sure we deliver the operational telescope to the moon. To do this, we choose to launch it into space with a rocket that has flown before. Why? Because, if you look at the history of new rockets, most are tragically unsuccessful on their first flight or flights. You won't want to risk your mission on an untried rocket, even if it saves 50 percent on launch costs. That's why new entrants in the space launch business, like SpaceX and Blue Origin, need to be self-funded for at least the first few flights.

The same argument can be made for most of the spacecraft support systems. The radio? Use what we've used before, even if it doesn't have the data rate you'd like. Getting the data back more slowly is more important than risking a new, higher-performing radio that might fail and not let you get any data back. The computer? Use the design that's been flown many times, even though it is based on an architecture that was available commercially before the smartphone was invented. Why? Because it works, and we've used it before. What about propulsion? Can we use one of the new high-performance electric or solar-sail propulsion systems to get us to our destination faster, using less power and less fuel? No. Too risky. These systems have only flown a couple of times in deep space, and we don't have enough data to really know their reliability. We will instead use a chemical rocket designed in the 1970s because we have flown hundreds of them, and we know they are reliable. Etc. Etc. Etc. In the end, the only new technology we often end up flying is the telescope. And, guess what? It usually works and the customer is happy. If the customer is happy, then they will come back with more business in the future, and the process will repeat itself. In the end, very little new technology is actually used.

Progress in this area is necessarily incremental. (For more information on the space applications of graphene, refer to chapter 9.)

Taking this back to graphene, and not even considering space applications, we should ask if our customer is willing to depart from what we all know works and take a chance on something new and better. Experience says they will not unless the payoff (profit) is potentially very high and worth the extra risk. In practical terms, this means that we are more likely to see new graphene-based products made by young or startup companies than by the existing market leaders in any given industry.

LEGAL

Graphene may have existed long before it's "discovery," but that does not mean that the methods of making it, or the myriad ways we might find to use it, are in the public domain and up for grabs. According to an analysis of patent applications by the Intellectual Property Office in the United Kingdom, the number of applications mentioning graphene has continued to rise every year (figure 8-1).[6]

It is difficult to see how anyone will be able to navigate this veritable sea of patents and not infringe on someone's legal claim in the process of making and selling a new product. Patent infringement lawsuits are not new. They go back to the earliest days of innovation; intellectual property ownership has been enshrined in Western legal systems for centuries, as have the various methods of protecting one's inventions from those who would unabashedly steal them. Consider the most well-known inventor in American lore, Thomas Edison.

Edison is credited with filing over one thousand patents in the United States, and at least that number again worldwide.[7] He is known for inventions such as the incandescent light bulb, the phonograph, and a motion picture camera. But there were many, many more patent applications that he or his employees filed that never saw the light of day. He sold patents to others to raise money to fund his laboratory and aggressively sued those who tried to market products that infringed on his patented ideas. I would use the term "defensive patents" to describe his patenting of ideas that he had no intention of turning into a product but thought someone else might—and that he could therefore get them to pay him for the right of doing so. One of his rivals, George Westinghouse, however, used patent law to establish himself as the preeminent competitor to Edison in power generation.

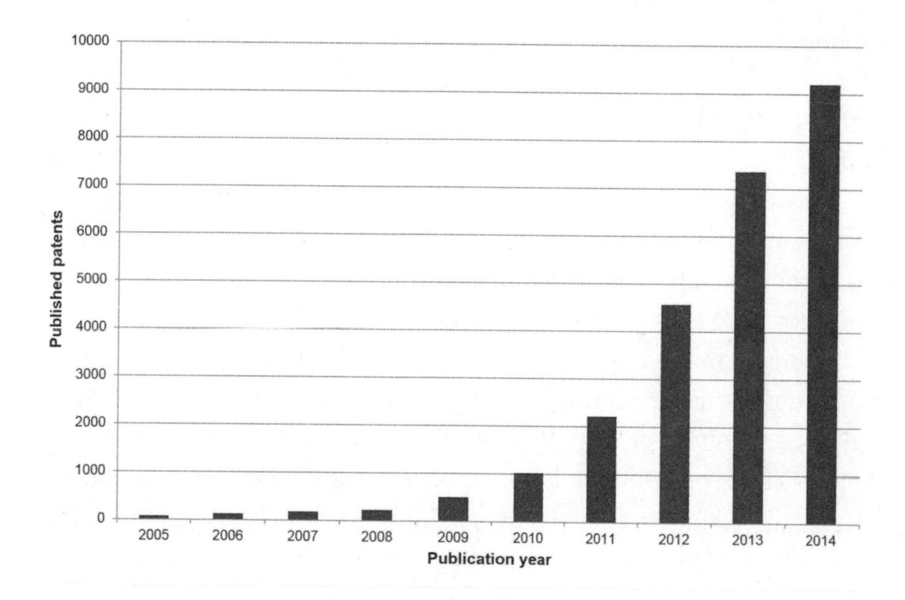

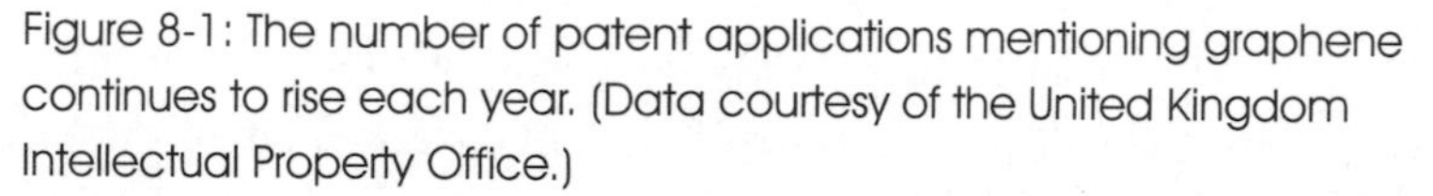
Figure 8-1: The number of patent applications mentioning graphene continues to rise each year. (Data courtesy of the United Kingdom Intellectual Property Office.)

Edison was convinced that direct current (DC) was the way to electrify the world. He demonstrated the wonders of DC power by using it and his incandescent light bulbs to light up whole city blocks. But DC power had its limitations. It couldn't be used long distances from where it was generated, a problem that still exists today. He knew this was a problem, so he hired a bright young inventor, Nikola Tesla, to solve it. Tesla did just that and proposed to Edison alternating current (AC), which would allow power transmission across vast distances with minimal loss. The details are sketchy, but Edison apparently dismissed the idea and fired the young Tesla. Tesla then filed his own patents—a smart move—as he tried to raise money to start his own electrification business. His invention caught the eye of George Westinghouse, who bought Tesla's patents and began building his own power-generating system. The rivalry between these two tech giants, Edison and Westinghouse, raged for years, with Westinghouse eventually prevailing, and our homes are powered by AC systems today.[8] The intellectual creativity war between Edison and Tesla is also legendary, with many still debating "who was the more inventive" today. The salient point is that Westinghouse didn't invent AC power, Tesla did. Westinghouse saw a good idea, bought the idea (by buying the patent), and funded the inventor to help make the patented idea a reality. This is ideally how the system is supposed to work. Sometimes, however, in our world of attorneys, it doesn't work so smoothly.

There is an ongoing legal feud between the two largest smart phone manufactures in the world, Samsung Electronics Co. and Apple Inc. At stake are enormous profits and control over the entire smartphone industry. Apple, which made the first smartphone, runs Apple's unique operating system and claims intellectual ownership for many design features due to patents they filed outlining them. For example, Apple patented the basic shape of the iPhone, its graphical user interface (apps anyone?), and other features. Samsung countersued, of course, claiming that Apple infringed on some of its own design features. Lawsuits were filed in American, South Korean, and German courts, as well as in many other countries. Both companies actively worked to get favorable rulings in one court or another to bolster their ownership cases globally. The suits, countersuits, verdicts, appeals, and new lawsuits continue. The primary reason they can continue is the sheer size of both companies—their vast profits feed their deep pockets to continue funding their legal teams. Had any of these patents been filed by a sole proprietor or a small business, there is almost no way they could prevail against the legal onslaught.

Let's go back to the graphene patent bonanza that is happening today. Universities, companies, research laboratories, and individuals are innovating new methods to manufacture, use, and modify graphene, as well as, in many cases, extrapolate its use in applications not yet even remotely practical today. Considering the historical patent infringement cases summarized above, this might, at first, seem to be a logical thing to do. You have an idea, and, to prevent people from stealing it, you patent it and hope you have the legal resources to enforce your patent when someone begins to infringe upon it. Unfortunately, this may make sense when your innovative product is about to hit the market, but it may not make as much sense in the realm of fundamental research and what we call "Idea Space." It is possible that all the patents, the infringement lawsuits, and the protracted legal battles will only serve to keep graphene-based products out of the marketplace much longer than would have otherwise been the case.

Until 1980, American universities really didn't patent many of their innovations. Research universities worked to advance human knowledge and educate students to be the next generation of innovators. After all, most universities are funded by tax payers, and why should the university or individuals working there be able to profit while working at the public expense? This all changed in 1980, when a new federal law allowed universities to attain ownership of patents arising from federally funded research. This law changed everything.[9]

Universities set up technology transfer offices to oversee the patenting of new ideas and to help spin them off to corporations via licensing agreements and partnerships. Instead of a discovery simply being written up and published in a journal, it is now assessed for commercial potential and possible enrichment of the university. According to the *New York Times*, federally funded research universities each year collect approximately $2 billion in licensing revenues and issue over four thousand patent licensing agreements.[10] Do you think this affects how universities decide to allocate their research dollars? You bet it does. It also complicates the legal morass surrounding the commercialization of products. Now you have federally funded universities licensing intellectual property that they discovered and patented from research funded by the public and then aggressively using their taxpayer-funded attorneys to help enforce these same patents against infringement. And, in the case of graphene, this may be a huge problem. Why? Many of the graphene applications being patented are, at this time, strictly notional. They are simply "ideas," without sufficient technology to make them real. In the past, these ideas would have been published, protection free, for all to see and assess. It wasn't until the actual widget based on the idea was invented that the patents would be filed and the protections thereby afforded put in place.

Graphene is subject to the same laws of supply and demand that govern the cost and availability of all other products in the global marketplace. For any of the revolutionary graphene-based products to succeed, they will have to overcome production challenges (quantity and cost), market inertia (cost of competing approaches and products), and legal wrangling (patents and intellectual property). With so much money at stake, the global race to overcome these challenges is widespread, well-funded, and unfolding at breakneck speed.

Part Four

WHAT'S NEXT?

Chapter 9

GRAPHENE IN SPACE!

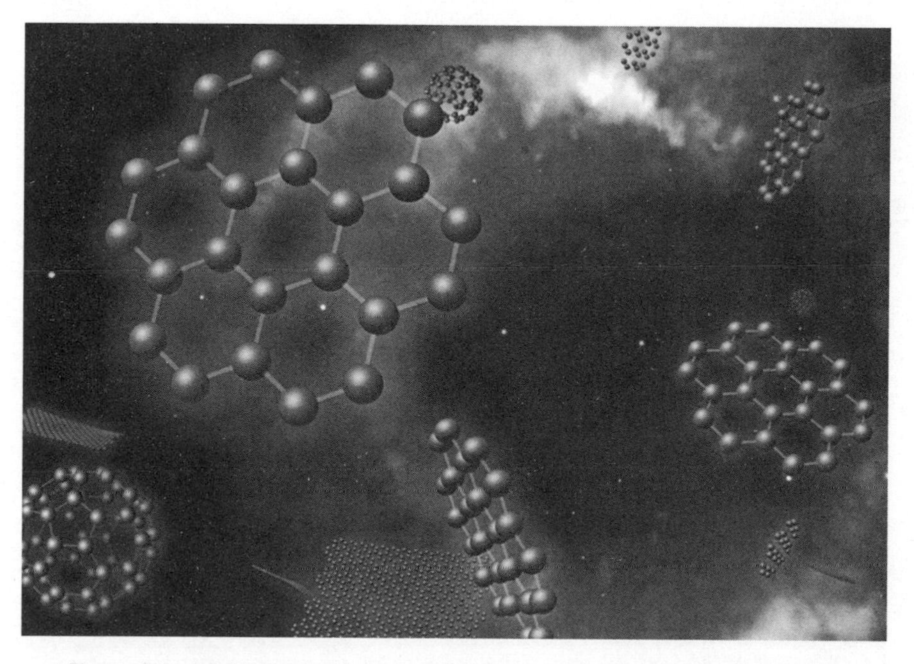

Figure 9-1: Artist's conception of naturally occurring graphene in space. (Image courtesy of NASA.)

NASA has detected naturally occurring graphene in space. While we have been puzzling over how to make and use graphene to help us explore space, nature made some out there for us to discover. In 2011, NASA's Spitzer Space Telescope, a sister of the Hubble Space Telescope that looks at the universe in infrared light instead of visible light, found what appears to be naturally produced graphene sheets among chemically related and also naturally-occurring buckyballs in the Magellanic Clouds, small companion galaxies that lie just outside of our own Milky Way galaxy (figure

9-1).[1] The graphene was found in various planetary nebulae within those galaxies, raising the possibility that naturally occurring graphene might have been present when our own solar system formed and could still be around today. Could these graphene sheets have formed through a stellar version of the Kansas State explosion experiment? We will now explore how graphene can be used to help us with our exploration of space, perhaps one day allowing human explorers to sample nature's naturally occurring deep-space graphene in person.

Space exploration is limited by many things, but one of the most critical is mass. The more the spacecraft weighs, the more difficult it is to get it from place to place—in cost and technical complexity. The reason is simple: Force = mass × acceleration, $F = ma$—Newton's Second Law, one of the most fundamental physics equations ever developed. Simply stated, it means that given a constant force, F, a given mass, m, will experience an acceleration, a. And since a mass has to be accelerated to travel from place to place, some sort of force needs to be applied to move it. And that force has to increase as the mass increases, or the acceleration will be too small and the mass won't be able to get anywhere very quickly. This is the problem that limits our exploration of space to only what is relatively nearby.

Believe it or not, modern spacecraft are not typically built from the latest and greatest wonder materials that scientists have developed in their laboratories. No, space mission designers are notoriously conservative in their approach and tend to select materials that have already flown in space many times, perhaps hundreds of times. Why? Because they're proven. People have built spacecraft from them before and flown them successfully to space, in Earth orbit, and beyond. This conservatism isn't borne from a lack of creativity but from economic necessity. Building something to fly in space is expensive, and those paying for the project don't typically want to accept too much risk.

Think about the problem from a rocket scientist's point of view. Someone is paying you to build a spacecraft to perform some mission. It could be a communications satellite that has to orbit the Earth for the next twenty-five or more years, relaying cable television signals or the internet all over the world, twenty-four hours per day, every day, without fail for the next quarter of a century. Any interruption in service will mean that millions of paying customers will be without service and looking for an alternative—costing your employer money. Or maybe your spacecraft is intended for a deep-space scientific mission to explore the moons of Neptune. In this case, the spacecraft will have to travel billions of kilometers through space to reach its destination, taking perhaps a decade just to get there, and then it must operate for years as it zooms past the various moons, studying them and relaying important scientific information back home.

In both of these cases, the "heart" of the spacecraft isn't its structure. No, the "important" part of the mission, including new technology, is in the payload. Whether it be a data transponder for the communications satellite or a high-resolution camera for the deep-space science mission, this is where the customer is expecting to accept his or her risk. The structure of the spacecraft just needs to hold the spacecraft together during all phases of the mission, and, hopefully, not pose any significant risk along the way. The spacecraft's structure has to survive three to five times the acceleration of Earth's gravity—hence three to five times its relative weight—during the launch into space aboard whatever rocket is taking it there and a range of atmospheric pressures ranging from zero (in space) to 1 atmosphere (on the launch pad). It has to be able to survive and distribute the heat when it is exposed to the energy of the sun, which is not attenuated by an atmosphere like it is for us here on Earth. No, in space, the full fury of our nearby ball of fusion-heated hydrogen plasma mercilessly bakes anything exposed to it. And it has to survive the opposite extreme, cold, when the spacecraft either enters the Earth's shadow and faces the near-absolute zero temperatures of deep space or it has to survive the cold directly as it traverses the distance between the planets near the edge of the solar system.

For these reasons, the space industry settled on two elemental materials many years ago, and it is very reluctant to change from them: aluminum and titanium. Titanium is strong, lightweight, and works well with extreme variations in temperature and pressure. Aluminum is inexpensive, lightweight, and can be easily manufactured, milled, and shaped in just about any machine shop on Earth. But, when compared with the composite materials now used in making nearly everything from cars to aircraft, titanium and aluminum might as well be lead. And that is a problem.

Nearly everything else that goes into a spacecraft has grown smaller and lighter-weight. For example, the electronics revolution has made the so-called avionics suite, the set of electronics that consists of the flight computer, the sensors that tell the spacecraft where it is located and how it is pointed, and the onboard radio for hearing commands from home and transmitting data to customers, smaller and much less massive. Just think of your cell phone and compare it with the computers of just a few years ago. That's the kind of miniaturization that has revolutionized the aerospace industry. But most spacecraft hulls are still made from variations of the same materials that were used in the 1950s and 1960s.

While the automobile and aircraft industries have embraced some of the lightweight composite materials described in chapter 2, many of which are made from carbon, the space industry is the notorious holdout. Carbon composites have made headway into the so-called secondary structures, those that strengthen or connect the primary materials from which the spacecraft is made, but few have been used to build the actual spacecraft itself. This may change with graphene.

Why might graphene succeed where other new materials have failed? Because graphene doesn't just offer the same performance as titanium or aluminum while using less mass, it offers dramatically superior performance with *much* less mass. As we have emphasized throughout the book, graphene also offers the possibility of having instruments and sensors integrated within the structural material itself, taking advantage of graphene's unique conductor and (hopefully) semiconductor properties, potentially eliminating the traditional idea of a "structure" altogether. Future spacecraft may be made with graphene in such a way that the distinctions between instruments, communications systems, sensors, and scientific payloads are completely impossible to discern.

Aside from the obvious benefits of graphene in reducing the overall spacecraft mass, strengthening it and making the structure lighter, graphene could also enhance or enable some novel, and highly promising, advanced space propulsion technologies, such as solar sails and electrodynamic tethers.

Figure 9-2: The Planetary Society's LightSail-1 captured this "selfie" during its Earth orbital flight in 2015. The spacecraft deployed a thirty-two square meter reflective solar sail. (Image courtesy of the Planetary Society.)

Solar sails are basically large, lightweight, reflective sails made from space-durable polyimides (plastics) and coated with something that reflects sunlight. As their name implies, as the sunlight reflects from the sail, they move—just like a sailboat moves when the wind reflects from its sail. Solar sails are typically very large because the pressure of sunlight, and the resulting force, is very, very small—on the order of the force equivalent of the weight on Earth of a quarter and a dime held in your hand for a sail the size of two football fields! Today, sails flown in space have areas of between one hundred and one thousand square meters and weigh a couple of kilograms. They are currently used to propel very small spacecraft (less than twenty-five kilograms, or about the weight of a sack of potatoes) within the inner solar system. Unfortunately, not many spacecraft weigh only twenty-five kilograms. If solar sails can be made lighter and stronger, then they can made to have larger areas, which reflect more light, produce more thrust, and can then carry much more massive spacecraft.

It is here the real utility of solar sails becomes clear: they do not require any fuel. As long as the sun shines, they produce thrust and can continue accelerating, never running out of "gas." To date, nearly every spacecraft mission flown has used some sort of rocket for propulsion. A rocket is basically any type of propulsion system that expels a propellant from one end to move the spacecraft in the opposite direction. There are chemical rockets, like those that are used to get from the surface of the Earth into space and from there to anywhere in the solar system. The problem is that rockets require fuel, lots of it, and they use it up very, very quickly. Consider the rockets SpaceX uses to launch satellites into orbit and send supplies to the International Space Station (ISS). Fully fueled, on the launch pad, the Falcon 9 rocket weighs about 505,000 kilograms. The rocket can insert about 11,000 kilograms into low Earth orbit. That means the mass of the payload is only about two percent of the overall rocket's weight. Two percent! The rest of the weight is allocated to the structure, the electronics (a very, very small part of the weight), and, most of all, the fuel required to get it into space. Get this—it takes only about eight minutes for the Falcon 9 to get from the surface of the Earth into Earth orbit, burning tons of fuel in the process. *Tons of fuel are used in less than ten minutes*.

The situation is no different with rockets used solely in space, to get from point A to point B, where A and B can be nearly anywhere in the solar system. The chemical propulsion systems used on nearly all of the deep-space exploration and science missions flown to date have had their total launch mass be approximately 50 percent fuel. For every kilogram of spacecraft or science, there had to be a kilogram of fuel added. And, like their Earth-to-space counterparts (i.e., the Falcon 9 rocket), most of this fuel was used in the first few minutes of flight. The spacecraft then coasts, without accelerating, for years—let me say that again—*years*—before reaching its destination.[2] Would it not be better to take advantage of a lower thrust system, like a solar sail, which continues to accelerate as long as the sun shines, using no fuel, to send spacecraft to their final destination? The answer is often yes, and that is why the technology is being developed.

Solar sails can enable spacecraft to orbit the sun's poles without requiring massive rockets or to make long trips out to Jupiter and back, taking advantage of the giant planet's mass for a gravity-assist maneuver—and adding years to the possible trip time of such missions. Solar sails can enable low-cost reconnaissance of near-Earth asteroids with small spacecraft—which NASA is planning with its Near-Earth Asteroid Scout mission.[3] They can be used to keep spacecraft constantly thrusting along the sun-Earth line to monitor and provide advanced warning of solar storms—an essential service for protecting and extending the lives of some very expensive spacecraft in high-Earth orbit.

Given that current solar sails are already thin (approximately half the thickness of a sheet of notebook paper) and lightweight (about twenty-five grams per square meter), making them thinner and lighter seems daunting. And it is imperative to maintain their robustness in the process; as you probably imagine, sails don't work well when they have holes in them. Materials that get this thin tend to tear or rip easily. Thinner sails made from today's state-of-the-art materials tend to get very fragile after extended missions and therefore become unusable. But we need sails that are larger in area, thinner, and less massive to achieve some of the impressive space missions enabled by solar sails like those envisioned for the exploration of nearby interstellar space. And graphene may just be the material needed. But there is a problem: graphene isn't naturally reflective.

In one of the Star Wars prequels, there was a scene involving a solar sail that, of course, got it all wrong. In the movie, a solar-sail propulsion system deploys from Count Dooku's space yacht. On screen, it looks awesome. There is only one problem. It is not exceptionally reflective—it is dark. We can build a solar sail today that is more efficient than this one, just by making it reflective. And, unfortunately, sails cannot take us into hyperspace—if such a thing even exists! Solar photon pressure, the force of sunlight that pushes a solar sail, will work if the photon is absorbed by the sail. This happens when a sail is dark. But that same photon can produce twice the thrust if it reflects from the sail. To do this, the sail has to be reflective and shiny, not dark. So, while Count Dooku's solar sail was impressive in size, it could have been half the size if he had bothered to add a reflective aluminum coating on the outside. The same will be true of graphene solar sails. They will need to be coated or doped with something to make them reflect more light than they do in their natural, absorptive state. But designers will have to be careful. Any coating they add will increase the weight of the sail and reduce its overall performance.

Now let's dream about how graphene solar sails might help us reach the stars. The Interstellar Probe is a science mission envisioned as the successor to Voyager. The twin Voyager spacecraft were launched by chemical rockets back in 1977 and today hold the record for being the most distant spacecraft from Earth. They have traveled more than 130 Astronomical Units, or approximately 149 million kilometers, and are leaving the solar system toward the stars at a speed of seventeen kilometers per second.[4] After over forty years of flight, they will soon die, as their plutonium power packs decay past the point where they can produce useful heat and electrical power. But what if we can launch a new spacecraft, one that travels at speeds five times faster than Voyager, so that we can explore more distant space and not have to wait until we're dead to analyze the science data? That is the challenge for the Interstellar Probe. And it is a challenge that can be met with a solar sail.

Analysis shows that a Voyager-class spacecraft, propelled by a solar sail that weighs one gram per square meter or less (compared to today's sails weighing twenty-five grams per square meter) with an area of at least 90,000 square meters (versus today's one hundred to one thousand square meters), then we can build and launch the Interstellar Probe and get data back within ten to twenty years of launch, from distances as great as three hundred Astronomical Units. Graphene can make this happen. Given its strength and very low mass, large sheets of graphene, coated with a reflective layer like aluminum or beryllium should do the trick—easily. Such sails would be as robust, or more robust, than today's sails. They would be larger and weigh considerably less. Graphene sails may actually enable us to go still further and send our first probe to another star.

The next step beyond solar sails are laser sails. As their name implies, high-energy lasers will replace the sun as the source of energy to propel them. Using a laser will allow much more concentrated light energy to be reflected from the same area of sail, thus producing significantly more thrust. There is an additional materials problem that arises when high-energy lasers are used: sail heating. Reflecting sunlight is one thing; reflecting thousands or hundreds of thousands of suns of equivalent energy from the same sail area is quite another. Without having a coating that reflects essentially all of the incident light energy, most known materials will simply melt from the absorbed (not reflected) heat. As an example, today's state-of-the-art solar-sail material, the one that is being flown in space by the Near Earth Asteroid Scout mission, has a reflectivity of about 0.93, where 1.0 is a perfect reflector. That is pretty good. But 0.93 reflectivity means it has an absorptivity of 0.07—it will absorb 7 percent of the light energy that strikes it.

Imagine that we build a sail that is not measured in square meters, but square kilometers. Think of a sail the size of Texas that is as thin as a single layer of graphene—one atomic layer. Now imagine that we deploy it close to the sun to take advantage of the plentiful sunlight there, which accelerates it much more rapidly than if it were to deploy near the Earth. As it flies by the Earth, and the sunlight intensity begins to drop with distance, we shine a powerful laser on it so that the sail continues accelerating. This laser is as powerful as the 2017 energy output of humanity over an hour, on the order of a few terawatts, and we are able to keep it shining on the sail as it departs the solar system and enters interstellar space. Such a sail might reach the nearest star in less than a few hundred years. Compare this with the 70,000 years it will take Voyager or other chemical-rocket-propelled spacecraft to go this distance and you can see what a revolution this will be.

Graphene may enable us to reach the stars.

Graphene just gets stranger all the time. *New Scientist* reported that researchers at Nankai University in Tianjin, China, built what they called a graphene sponge made from combining several layers of crumpled graphene oxide.[5] When they shined a laser on the sponge, it moved. Now, at first glance, this should not have happened. Recall that the force of light is very, very small and is easily swamped by other forces acting on an object here on Earth—especially gravity. Shining a laser on a solar sail in the laboratory results in no perceptible movement—except to the most sensitive of instruments. But this group found that the graphene sponge moved several tens of centimeters when the light shone upon it. The most likely alternative explanation was that the laser vaporized part of the sponge, boiling some of the material off, which might cause the sponge to move in the opposite direction.[6] Only when they looked closely at the surface interaction, that wasn't happening.

Another theory, which looks like it might explain the motion on a gross level, is that the laser light ionized the material, causing a buildup of electrons that then flew off the sail material causing the sail to recoil (move). If that is the case, then there is the question as to why the electrons all flew off in a single direction and not isotropically (in all directions), which would have resulted in no net motion.

So, why is this important? If the sail is moving because of electron emission and not some odd thermal effect (heating the air, etc.), then, in space, it could become a propulsion system that has the advantages of a highly efficient, light-reflecting solar sail, along with being an electron-emitting rocket. Together, the two physical phenomena might allow a spacecraft attached to the graphene sail to fly throughout the solar system quickly, using almost no fuel.

Let us take a break from solar sails and dreams of reaching Alpha Centauri by 2030 to talk about another space application of graphene—as the structure for a space elevator. For those not familiar with the concept, a space elevator is simply an elevator that takes you to space. Humans have dreamed of making structures that reach far into the sky since the beginning of recorded history. Consider the biblical story of the Tower of Babel, as told in the book of Genesis:

> And they said, Go to, let us build us a city and a tower, whose top may reach unto heaven; and let us make us a name, lest we be scattered abroad upon the face of the whole earth.[7]

Modern humans build fantastically large structures, and at the time this goes to press, the tallest building in the world is the 828-meter-tall Burj Khalifa in Dubai. (As a comparison, the US Empire State Building is merely 443 meters tall.) Now, imagine a building or tower that reaches altitudes greater than 42,000 kilometers, and imagine further that it has an elevator that you can ride to the top. In theory, one could send people, cargo, or spacecraft up this elevator directly into space. The attractive notion of this structure is that such trips would only require the recurring cost of electricity used. No rocket. No rocket fuel. Inexpensive and simple. Well, not so simple . . .

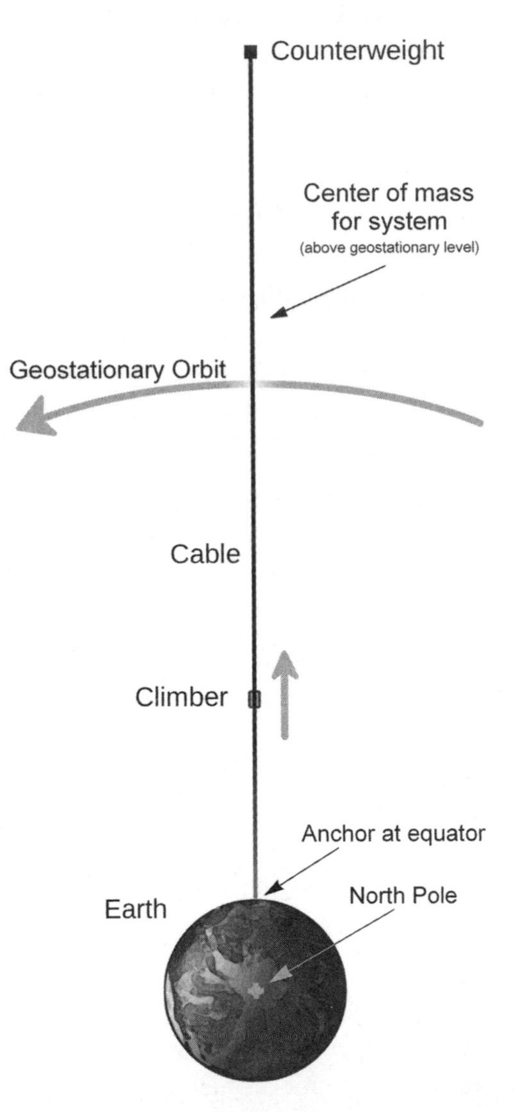

Figure 9-3: Artist's concept of a space elevator built upward from the Earth's equator to reach above geostationary orbit. (Image courtesy of NASA.)

How in the world, pun intended, might we build a space elevator? Can it be done? There have been many conceptual designs for futuristic space elevators, and most call for a cable with one end attached to the Earth's surface near the equator with the other end in space. The cable is kept vertical by putting it under tension, in a manner similar to how a yo-yo string, when spun over your head, is kept in tension and doesn't turn into a limp noodle—centrifugal force. The end close to the Earth is kept under tension by the planet's relatively strong gravity, and the rest of the cable is kept under tension by the centrifugal force generated by attaching a small asteroid to the tip of the tether that resides in space (which is rotating, like our yo-yo, since the Earth rotates). Voila! We have our vertical tower, extending from the surface of the Earth into space.

We should get real at this point. We have never built anything of this scale, and the analysis indicates that, in order for it to work, if we can find a way to construct it, it would have to be made from a material stronger and lighter than any previously known material. The elevator would have to be strong enough to sustain its own weight as well as the tension placed on the elevator by the asteroid anchor on the top end. Ideally, it would be electrically conductive, so you could actually use the structural material for the elevator as part of the power system and you won't have to worry about constructing a 22,000-mile power cord that cannot sustain its own weight. Based on these requirements, is it starting to sound like something familiar might just be an option for making a space elevator possible? Graphene, or its cousins, carbon nanotubes, are theoretically the only materials known today that might enable this.

The astute reader might notice the weasel word I used in the preceding sentence—"theoretical." I thought graphene was real? Why do you call it theoretical? The answer is simple: Until we know how to make long—extremely long—wires from graphene, many of its macroscale applications will remain "theoretical." To really design or build something, or plan to do so, you need to know that the materials from which it will be constructed are real and meet the design requirements. With regard to making graphene cables many thousands of kilometers long, the verdict is still out. To get a sense of the scale of the problem, consider that the Earth's circumference is approximately 40,000 kilometers—about the same as the height of the space elevator.

Here is an interesting but important side note about sending things into space using a space elevator. For something to orbit the Earth, like a satellite, it must be moving relatively fast (about 28,000 kilometers per hour) around the Earth. That is, it must have sideways speed, not just vertical speed, to be in orbit. This means that anything sent to space by the elevator would have to ride it nearly all the way to the top to be in orbit around the Earth and not fall back toward the ground. If something is released from the elevator at any altitude below geostationary orbit, then it will fall back toward the Earth due to the Earth's gravity. Mental note: Don't stand next to the space elevator lest something fall on your head and kill you . . .

In chapter 4, we described additive manufacturing (AM) and 3-D printing using graphene. NASA considers AM to be a critical element in support of human spaceflight and is investing in the technology in a big way. Using graphene will only make it more capable.

NASA flew AM systems on the International Space Station (ISS) to test their operation in the weightless environment of space because they see the technology as a needed one for long-term space exploration. When human space missions are planned today, a great deal of launch mass is set aside for spare parts. Who would want to be on the way to Mars on a two-to-three-year round-trip mission and have some critical part break without there being a spare to replace it? Statistically, not every critical part is going to break, but it is almost inevitable that at least one of them will. How do you plan for that? By bringing spare parts for the most critical systems with you on the trip. This means that, in addition to launching the fully functional systems needed to support a crew on a mission, NASA needs to launch a repository of spare parts to be accessed as needed for in-flight repairs. Most of these space parts will never be used, but they are needed "just in case."

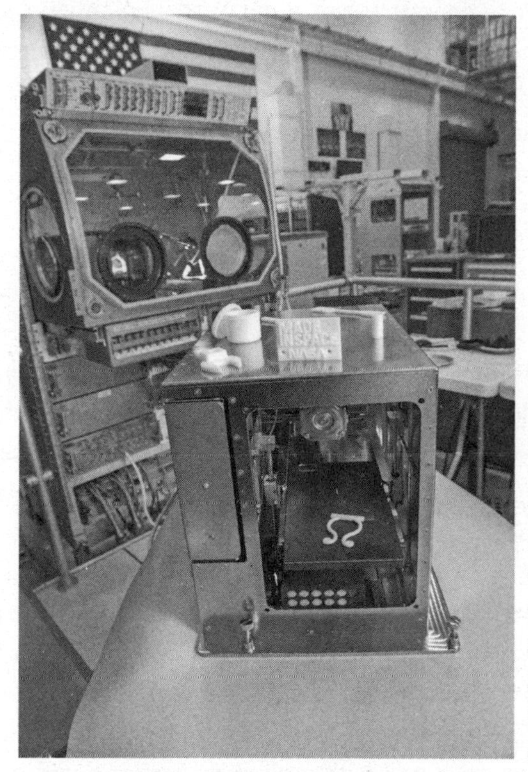

Figure 9-4: This first-generation, space-qualified 3-D printer, with the Microgravity Science Glovebox Engineering Unit in the background, was flown and tested in space aboard the International Space Station. (Image courtesy of NASA.)

The problem with this spare part strategy is that it can dramatically increase the cost and complexity of the mission. More spare parts mean more storage space and, most significantly, more mass. More mass requires additional fuel, which adds yet more mass to the overall system. With launch costs ranging from $3,000–$10,000 per kilogram, bringing all these spare parts and the added mass to store and propel them can be costly indeed.

What if you don't need to bring all these spare parts, but instead bring a 3-D printer, raw printer material, and the manufacturing plans for all the possible spare parts that might be needed? In this scenario, you can dramatically reduce the mass allocated to spare parts, reducing the amount of "stuff" you have to send to space—saving money and complexity at the same time. This would also give the astronauts and mission planners more flexibility. If some device or part is needed that was completely unanticipated, the plans for making it could be sent to the spaceship by radio from Earth, where the engineering design talent is essentially unlimited.

For science fiction fans, particularly those who watched *Star Trek*, this may start sounding familiar—we are in the infant stages in the development of a "replicator," capable of making whatever is needed by our intrepid crew as they explore the solar system and beyond. So where does graphene fit into this vision? Everywhere. For all the reasons we described, from its material strength, electrical conductivity, and, when properly doped, semiconductor properties, to its curious properties that can be exploited for water filtration, electrical power generation, and storage, having a 3-D printer capable of working with graphene will open all sorts of possibilities:

- Making habitats on the surface of the moon or Mars using large-scale 3-D printers that mix the local dirt, or *regolith*, with graphene to make the habitats stronger and more survivable in the harsh environments of either planet.
- Printing these surface structures with embedded, printed sensors to monitor the exterior and interior environments of the habitat.
- Making critical life-support systems *in situ*, adapting them to the unique characteristics of the particular landing site rather than having to either first design them on Earth to survive in a particular location, limiting your exploration options, or making them robust enough to place anywhere, increasing their mass and complexity. Instead, design and build the life-support systems you need, when you need them, and in the environment in which they are needed.

Finally, graphene-enhanced sensors can be used to manufacture science instruments for space applications. Graphene's properties are attractive for terrestrial applications and also ignite interest in the space science community—where scientists are always looking for ways to make their instruments smaller, less massive, and more energy efficient.

An example of such a benefit would be the sensors used to measure atomic oxygen concentrations in low-Earth orbit (LEO). Oxygen atoms are usually diatomic, meaning they bond to each other in pairs: O_2. Rarely on Earth do you find a single oxygen atom standing alone. But in LEO, which is not a pure vacuum, you have the energy of the un to split apart O_2 into single oxygen atoms. These atoms can wreak havoc on materials in space over time. Many materials are literally etched away due to their continuous exposure to atomic oxygen, causing them to break or stop functioning. Today's atomic oxygen sensors are not exactly massive, but work conducted at the NASA Goddard Space Flight Center points to a new generation of graphene-enabled sensors embedded into spacecraft structures that could detect atomic oxygen as well as the concentration of other neutral atoms. These sensors could then be modified for use on missions throughout the solar system, allowing scientists to characterize the atmospheres of other planets using only a small fraction of the mass and power currently required.

Graphene has the potential to revolutionize space exploration and enable ambitious, world-changing missions, now considered to exist solidly only in the realm of science fiction.

Chapter 10

GRAPHENE CYBERNETIC ORGANISMS

Cyborg is short for cybernetic organism, a term used to describe someone or something that is purposefully modified to become more than biological. The modification can be to correct for some biological deficiency; to augment a biological inadequacy, real or perceived; or to help it adapt for survival in some new environment. There are sure to be other reasons to make biological modifications, but it is these three that are most often cited. These augmentations can be mechanical, electrical, or biological, and they range from the complex and science fictional to the more mundane.

Fans of *Star Trek: The Next Generation* will immediately envision the ultimate cybernetic organism, the Borg, and perhaps the most famous Borg of all, Locutus (aka, Captain Jean-Luc Picard). The Borg are a collection of fictional alien races that have been turned into cybernetic organisms functioning in a hive mind called "the Collective," losing their individuality in the process. In this science fiction universe, when humanity encounters the Borg, they are immediately at risk of being overtaken and turned into cyborgs like the Borg (hence the name).

This notion is reinforced by other science fiction books and movies, including the wildly popular *Doctor Who* franchise. Who hasn't shivered when the good Doctor encounters the Cybermen, a race of humanoids that have become more robotic than human. According to the series, the Cybermen began like us and then began experimenting with implanting more and more artificial parts into their bodies as their technology allowed. And, following the same logic that made the Borg such great villains, the Cybermen, too, lost their humanity and became more and more machine-like. Other organisms have been imagined as progressing in the opposite direction—*Battlestar Galactica*'s Cylons and *Blade Runner*'s Replicants started out as AI and became more humanlike, adapting artificial biology to their needs.

The takeaway is that Borg, Cybermen, and their ilk are no longer human (or never were, despite appearing so) and are therefore evil. The moral of the story is this: We will lose our humanity if we become cyborgs. The cultural belief is that there is something unnatural (and therefore inherently evil) about machine-augmented biology. Much of the appeal behind today's "all-natural ingredients" movement stems from consumers' discomfort with modern human-developed ingredients within food and personal care items. Evil overlord Terminators are a fair warning of technology gone amok, but . . .

Most of us are already cyborgs, like it or not.

For example, many of us wear eyeglasses or contacts to correct for some vision inadequacy. Using an optical lens to correct vision dates to the thirteenth century. The earliest corrective lenses were used by monks and scholars and were held in front of the eyes when needed. In the eighteenth century, modern eyeglasses began to take shape, literally, as frames that rested on the nose and ears were invented. As most American school children are taught, Benjamin Franklin invented bifocals later in that century. Thirteenth-century monks had no way of knowing that they were cyborgs; science fiction was not invented as a literary genre until Johannes Kepler's *Somnium* was published in 1634. Modern vision correction takes the form of laser eye surgery, or Lasik, and cataract surgery. Cataract surgery is performed when your eye's lens develops a cataract, a clouding that makes objects look blurry or hazy. During the surgery, the cloudy natural lens is removed and then replaced with a clear artificial lens. That's right. Grampa, the one who says he can't figure out how to work the new smartphone you bought him, is a cyborg.

I (author Johnson) have a family member afflicted with type 1 diabetes. The diabetic pancreas produces little to no insulin because the body's immune system has destroyed the insulin-producing cells within it. Without treatment, the consequence is death. Those diagnosed with type 1 diabetes must inject insulin several times every day or continually infuse insulin through a pump, as well as manage their diet and exercise habits. Insulin therapy has a long and interesting history in and of itself, but, in brief, it all began in the early twentieth century with pig and cow insulin being used for the first time to save human lives. Today, most insulin is made from bacteria or yeast, using recombinant DNA technology. Basically, a human gene is inserted into the genetic material of a common bacterium or yeast. This recombinant microorganism then produces the insulin that is used to keep people with type 1 diabetes alive. (Yes, the yeast is also, technically, cyborg.) In the case of my family member, the insulin is released continuously into the bloodstream thanks to an electromechanical device that is semipermanently attached to their body for that purpose, an insulin pump. This family member also wears eyeglasses—a multifaceted cyborg.

These examples help us understand that discussions of creating cybernetic organisms is not purely science fiction and that cybernetic organisms are not necessarily evil or something to be avoided. In fact, like most aspects of human technological innovation, cybernetic augmentation is not inherently good nor evil. Only its uses and purposes can be judged in moral terms.

So how does this relate to graphene?

According to a paper by Emiliano Lepore of the University of Trento, in Italy, he and his research team did one such "what if" experiment by spraying a group of spiders with a mixture of water, carbon nanotubes, and 300-nanometer-wide graphene particles.[1] They then measured the strength of the silk the spiders produced and compared it with unsprayed spiders. Incredibly, by doing something as simple as giving the spiders graphene-laced water, they could alter the strength of the silk dramatically—making it more than three times stronger than natural spider silk, stronger than Kevlar, and among the most mechanically robust materials ever produced.

Figure 10-1: Researchers found that spiders fed graphene spin stronger webs. (Image courtesy of Christian Michel.)

Of course, the experiment was not without side effects. First, only some of the treated spiders actually produced this "super silk." Others produced normal silk, and a few actually spun silk that was well below the average in terms of quality. And—which is worth taking as a cautionary note to those of you ready to supersaturate your favorite pet with graphene water—some of the them simply died.

Now, think back to the summary above about the use of recombinant DNA to produce artificial insulin. This process is not done haphazardly. There are approximately three million Americans, and perhaps as many as seventy million people globally, with type 1 diabetes. Producing enough insulin to keep them alive and healthy is a major undertaking and is accomplished on an industrial scale. It is not difficult to imagine applying insulin-making methods to graphene production by taking Lepore's discovery, improving upon it by perhaps identifying the optimum way to introduce graphene into the silk-making process using recombinant DNA, and turning it, too, into a mass-production effort. This certainly sounds like a simpler approach to making some forms of graphene than Chemical Vapor Deposition, using industrial strength acids, or hiring thousands of people to isolate it from pencil lead with tape. Many of the applications discussed previously in this book, particularly those in chapter 5, might one day be enabled by the spinning of silk by industrialized spiders. Now let's transition from spider silk to other fibers—those used in the clothes we wear.

Scientists have teamed up with the fashion industry to develop clothing embedded with graphene-enhanced electronics that light up in response to the wearer's breathing patterns or other physiological changes. These self-powered sensors can be connected with low-power LED lights and programmed to change color as your breathing pattern changes. If you are resting and not breathing too deeply, they might light up as blue. As you walk and begin moderate exercise, they might emit a green color. During your morning job, they might start flashing yellow or orange—take your pick. Combine this with a low-power Bluetooth connection to your next-generation smartphone, smart watch, or other health-monitoring device, and you begin to get full-body physiological diagnostic information.

While this might be entertaining for those information junkies who simply have to have the latest fitness gadget, it might be critical for the medical community to help assess the condition of those at risk from various medical conditions. Elderly patients with heart disease, patients with compromised breathing from COPD (chronic obstructive pulmonary disease), tuberculosis, or other respiratory ailments might be able to have a real-time link to a smart system in the cloud to help them self-monitor their pace and give them warnings when they reach their individualized health limits.

More likely, at least for some of us, we will hear our personal artificial intelligence exercise coaches telling us to "Pick up the pace! At this rate, you won't meet your goals for the day! Way to go!"

Figure 10-2: Fashion trends come and go. What was in style one hundred years ago looks strange to the modern reader, just as today's clothing might appear to someone one hundred years from now. (Image taken from page 248 of "London [illustrated]. A complete guide to the leading hotels, places of amusement . . . Also a directory . . . of first-class reliable houses in the various branches of trade." Image courtesy of the British Library.)

It is not too big of a leap to imagine that different types of sensors might be embedded in our clothing to measure more than just our respiration rate. What about blood oxygenation levels? Blood sugar? Or perhaps the presence of various infectious diseases. Siri could tell us, "You have just been exposed to the viral meningitis. Please see your medical professional immediately to take countermeasures!"

Let's leap from graphene-augmented clothing to graphene-augmented medicine and humans and consider the possibilities.

Are you concerned about your health when you learn that several of your ancestors suffered from the same, or similar, maladies? Does a particular form of cancer run through your family? Would you like to know if you carry the genes that might put your future and planned offspring at risk for color blindness, diabetes, or autism? Graphene-enhanced sensors might be able to help.

Consider this: scientists in India and Japan are working to develop graphene-based transistors to detect harmful genes.[2] These sensors work through a process called DNA hybridization, which occurs when a "probe DNA" combines with its complementary "target DNA." Electrical properties in the probe change when this combination, or hybridization, occurs. This process is possible without graphene, but it requires several intermediary steps and the use of additional materials and processes. In other words, it is complicated. Use of graphene allows the researchers to skip these intermediary steps and improve the overall performance of the technique.

But what about those difficult-to-detect diseases that are all too often discovered too late for treatment? We have all had friends or family afflicted by the scourge that is cancer. If they were lucky, the tumor was detected early, before it had either grown too large or spread too widely. The odds of surviving cancer are dramatically improved if it is detected early. Unfortunately, since cancer cells are so similar to ordinary cells, our immune systems do not effectively combat them, and the they often go undetected until is too late. The spouse of a colleague was recently diagnosed with late-stage cancer and was dead within two weeks of diagnosis. She had been afflicted with it for many months before any symptoms were noticed, and by then it was too late.

Graphene won't likely be the "magic bullet" for cancer treatment, but it might be able to help with early detection. It will become one more important tool in the doctor's arsenal against cancer's detection and treatment. The reason is graphene's sensitivity to charge or any physical contact or presence on its surface. Recall that graphene is essentially a single atom-thick matrix of carbon atoms all lying in the same plane. It is extremely electrically conductive, and small changes in that conductivity, caused by any sort of surface contact, can be easily measured. Think of a thin layer of water flowing across a smooth surface and then introduce a rock somewhere in its path. The turbulence produced by the rock is immediately noticeable. The smooth surface is analogous to the single-layer sheet of graphene, the water is the electrical current, and the rock is an atom in contact with the graphene that is somehow "different." The degree with which the water flow, or electrical current, is changed, is indicative of the type of rock introduced. When a normal biological cell is in contact with an electrical graphene sensor, there is characteristic way the flow of water, or electrons, is disturbed that can be remotely measured. If the cell is cancerous, the flow is interrupted in a different way, or pattern, that can be detected using a technique called Raman spectroscopy. It seems that cancer cells tend to be much more active than normal cells (they are, after all, growing out of control—which is what makes them dangerous), and therefore they have a higher overall negative charge. It is this small charge difference that can be easily detected by graphene sensors.

The steps required to take this measurement approach from the laboratory environment to your local clinic are not clear. And for such sensors to act as an effective early-detection technique, they will have to make the transition from the clinic to our everyday lives. The goal is to find these cancer cells before they spread, out of control, through our bodies. Making them easier to detect using graphene is only the first step in this process.

What else can graphene-based sensors more easily detect? Going briefly back to the topic of diabetes, it is important to know that people with type 1 diabetes must accurately and regularly monitor their blood-sugar levels to know how much insulin to administer. If they inject too little, their blood-sugar levels remain high and, over time, damage is done to their circulatory system by large, sugar-laden blood cells rampaging through their capillaries and arteries. If they inject too much, their blood-sugar levels can rapidly drop way too low, causing unconsciousness or even death. Since the brain directly uses blood sugar for energy, the effects of low-blood sugar are felt there almost immediately. This tightrope of keeping blood sugar at the right level is a daily chore that people with type 1 diabetes must deal with at all times.

To accurately measure blood-glucose levels currently requires a drop of blood and a blood glucose meter. To get the blood drop, people usually have a small needle to prick their finger for testing. Under the best of circumstances, people with this disease must prick their fingers to test their blood sugar levels at least eight to ten times each day. Every day, of every week, of every month, of every year. You can imagine how tedious, inconvenient, and painful this must be. Surely there is a better way?

Scientists have found that blood-sugar levels can also be measured through analysis of tears. The amount of tear moisture to be tested is considerably less than a blood drop, and fortunately, very small sensors have been made to do this. Various companies have looked into placing these sensors into contact lenses that people with diabetes would wear as a substitute for continuous finger pricking and blood-sugar testing. The problem is that these contacts are very primitive, typically much larger and heavier than regular contact lenses, and they tend to cause dry eyes. This is where graphene comes in.

In chapter 6 we discussed how graphene oxide layered sheets can be used as filters for cleaning contaminated water. By layering two more sheets of graphene in the proper orientation, even water can be stopped, and you then have a nearly perfect moisture barrier. Combine this with the fact graphene is extremely lightweight and strong, and that it can also absorb electromagnetic energy (which, among other things, can be visible or ultraviolet light) and dissipate that energy as heat, and you have a material that might be a good candidate for use in blood-glucose monitoring contact lenses. The hypothetical graphene-based lens is strong, lightweight, protects the eye from damaging ultraviolet light, retains moisture to alleviate drying of the eyes, and gives the wearer information about their blood-glucose levels so they don't have to prick their fingers nearly as often to maintain good blood-sugar control. All in all, it sounds like a winner. Next, we will go a few inches farther in than the eye and look at applications of graphene within the human brain.

A group of researchers believes they have found a way to use graphene to make an improved interface between the neurons in the brain and the external world. (Their study used mouse brains, but that's normal. Mouse studies are often precursors to those performed on humans.) A team of researchers from the University of Trieste in Italy and Cambridge University created an interface between graphene and neurons that didn't damage the neurons in the process—a problem that has plagued previous attempts using other materials, which always resulted in a degradation of the neuron's ability to function.[3] The performance of previous implanted electrodes, typically made from tungsten or silicon, also degraded over time.

If these results are reproducible in humans, then we may not be too far away from graphene-based sensors measuring the brain's electrical impulses and correlating them with the subject's desired actions. Once this code has been cracked, the medical applications are plentiful. Artificial limbs for amputees or those suffering from paralysis might then be controlled by thought alone—dramatically improving quality of life. People suffering from Parkinson's or other neuromuscular diseases might have a new therapy to help them overcome the debilitating aspects of their affliction. Those whose eyesight is damaged or degraded might receive mechanical eyes that tie directly into the brain, restoring their ability to see.

The next logical step for this technology is intentional human augmentation. Can this neuron/graphene/electronic brain interface be used to improve the human body beyond normal biological limits? Imagine fighter pilots controlling functions of their aircraft by thought alone. Imagine soldiers on the battlefield equipped with artificially strong mechanical exoskeletons that move as easily as the soldiers' biological limbs due to the graphene-enhanced brain/computer connection—and with five to ten times the strength or speed.

It is not a stretch to imagine how this would work in one of the most challenging and complex battlefields of the twenty-first century—urban warfare. As the world has sadly seen play out in Syria and Iraq, today's soldiers are fighting and dying as they go from house to house trying to root out the enemy from among civilians within the confines of narrow city streets. Today's soldiers wear heavy body armor that slows them down and is only partially effective in providing protection. Ideally, in this environment the soldier wouldn't have to open and walk through a door, providing an obvious target for the enemy. Instead, the soldier could get a running start using his or her graphene strength-augmented legs, powered by lightweight and long-lived graphene-enhanced batteries or supercapacitors, burst directly through the wall taking advantage of the damage-resistant properties of the graphene-enhanced exoskeleton, and even take direct fire from the enemy in the room before completing the mission. If graphene-enhanced surfaces can resist damage during a hurricane or tornado, then they might be able to sustain direct hits from bullets and fully protect the soldier within.

This isn't a new idea. The military has been considering developing such *Iron Man* suits for decades and, until now, the results were not promising. An example of note is Project Hardiman in the late 1960s and early 1970s. General Electric, under contract to the US government, attempted to make a powered exoskeleton, which would have been similar to what Ripley wore in the movie *Alien*. The project came up with a test rig that weighed several hundred kilograms—far too much for a soldier to effectively carry into battle—which was largely uncontrollable. We've come a long way with computer control and miniaturization, as well as materials science, since that time. In 2015, the US military began testing the Tactical Assault Light Operator Suit (TALOS). TALOS uses the latest lightweight materials and state-of-the-art microcontrollers to make the exoskeleton more controllable and soldier-friendly. Researchers are also taking a more fabric-oriented approach (instead of rigid exoskeleton frames like previous systems tried to use). These not-yet-graphene-optimized suits now weigh only a few tens of kilograms and require only a few laptop-equivalent battery packs to operate. While that is a huge step forward, it is still not completely practical, as anyone who has had to walk very far wearing their winter clothes and carrying a laptop computer will attest. Graphene-enhanced components many just be the next technological step toward actual wearability with reasonable protection provided to the soldier and practical lifetimes between replacing the batteries or recharging (thanks to the graphene supercapacitor batteries). It will only be a matter of time before the suit's control system is connected directly to the soldier's brain to allow the suit to be an extension of the body instead of something that must be consciously controlled. Think of the difference between walking, which you do without thinking, and driving a car with a manual transmission, which requires nearly constant thought.

If the connection works one way, neuron to interface to outside world, then can it work in reverse? Could such implants be used in patients with severe burns to turn off the pain receptors while they heal? Could graphene-enhanced brain implants be selectively used to stimulate learning, improve memory, or help us learn to calm our most irrational fears (fear of flying, fear of heights, claustrophobia, etc.)? We don't know the answer to this yet, but researchers are working on it.

We know from functional Magnetic Resonance Imaging (fMRI) studies that different regions of the brain are active at different times, depending upon what a person is experiencing.[4] For example, certain regions of the human brain are stimulated when we see a familiar face versus one that is unfamiliar. When we commit something to long-term memory, especially when the "something" is associated with an intense emotional experience, we use the part of the brain known as the amygdala. When we sleep, the entire brain seems to be active, from the brain stem to the cortex. The cortex, which is usually associated with our sense of sight, is especially active and is thought to be the source of the storylines of our dreams.

It isn't a huge leap to imagine having graphene-augmented implants inserted into these brain regions to induce certain types of dreams, to make us more capable of learning, or to offset the effects of dementia. Of course, if this field goes the way of the internet, which allows on-demand access to the world's repository of knowledge and higher learning yet has porn as the number-one item being searched, someone will undoubtedly figure out how to stimulate the pleasure centers of the brain, providing orgasms "on demand." These same fMRI studies that are helping us unlock the secrets behind rational thought, about where in the brain we perform critical thinking, and where our creative genius is first sparked, also are being used to tell us which parts of the brain are active during sex. It seems that during an orgasm, the brain region behind the left eye (lateral orbitofrontal cortex) shuts down. For what it is worth, this is also the region believed to control our rational behavior. Hmm, go figure. . . .

Let's depart from our baser instincts, get more firmly into the realm of science fiction, and imagine training your brain to control systems that have no human body analogs: a ship's rudder or engine system; tens, hundreds, or thousands of drones flying in formation; or perhaps a network of cameras monitoring a city. A few years ago, there was a science fiction story about a man who died and woke up as a spaceship. His eyes were the cameras that monitored the inside and outside of the ship. His sense of temperature was the internal temperature of the spacecraft: cold feet meant the outer laboratory section temperature was low; sweating meant that the greenhouse was simulating midday summer sun. His arm flexing was the robotic arm used to load supplies from a visiting cargo ship into himself. His heart rate was an indication of how well the ship's propulsion system was functioning. You get the idea. Will graphene, combined with breakthroughs in biology and brain science, make this possible? Who knows, but it certainly seems like it ought to be—someday.

What if we think about this from the other direction? Can we use graphene-enhanced, living, biological organisms to improve our mechanical systems? Researchers at the University of Illinois at Chicago (UIC) believe so.[5] There, they created a nanoscale biomicrorobot that responds electrically to changes in its environment. To do this, they used a relatively benign bacterial spore that naturally responds to changes in humidity by either expanding, when water is present, or contacting, when it is not. They attached on each side of it a small bit of graphene and attached electrodes to the bits. As the spore shrinks, the graphene bits come closer together, increasing their conductivity, which can be measured by the electrodes. Given the organism's extreme sensitivity to changes in humidity, the response time of this new bioelectronic sensor is at least ten times greater than its purely mechanical cousins. Any mechanical, biological, or other process that is highly humidity dependent would benefit from the increased responsiveness provided by this smallest of cyborgs.

Before we get carried away about injecting ourselves with graphene, or even before we begin to mass produce it, we need to better understand how graphene interacts with the environment and us. In chapter 4 we looked in considerable detail at some of the potential biological and environmental effects of graphene. Here we will mention a few of the known health effects. Scientists at Brown University performed a study of graphene to look at its effects on human cells, and the results were alarming.[6] It seems that our planar, superstrong material is so strong that it easily punctured cell membranes in various human organs that are likely to encounter it: skin, lung, and immune cells to which it was exposed. Ouch!

It seems that if tiny bits of our friend graphene are inhaled into the lungs, they might just remain there, since there is no likely way for them to be broken down and removed. Remember, graphene is strong and durable; that is the reason we believe it will be so useful. If it lodges in the lungs, then graphene will act just like asbestos and other particles, causing the body to trigger an inflammatory reaction. One would think the immune system would send a few white blood cells to envelope the graphene and "take it out." Unfortunately, this doesn't seem to occur because of the average size of a graphene nanoparticle. They are simply too large for the immune system to deal with.

If these results are accurate, then those who work with graphene in its pure form must take precautions to protect themselves and to ensure that the material is not introduced into the environment willy-nilly. Most of us recall the wonder material that was asbestos, only to later learn that those exposed to it became ill with asbestosis and mesothelioma. Remember, we don't know (yet) if people will get this kind of exposure since we aren't really (yet) mass producing graphene. And it is likely that lessons learned from our asbestos-laden history will guide OSHA and other oversight agencies to come up with safe handling techniques to minimize these risks.

Recalling our discussion of plastic as a technological and societal disruption in chapter 7, we should be mindful of the massive and unintentional consequences arising from our use of that twentieth-century "wonder material." Most commercial plastics, including those plastic bottles so many of us buy for drinking water, can take hundreds, if not thousands, of years to decompose. Think about that when you casually toss your next empty water bottle into the trash instead of the recycling bin. Unfortunately, not many of us are recycling.

In 2014, global plastic production exceeded 300 million tons, with at least 8 million tons going directly into the oceans each year. And while the plastics in the ocean don't biodegrade, they do decompose into tiny plastic particles, pellets that can readily be ingested by fish swimming through contaminated water. Since the pellets aren't digested, they accumulate in the bodies of the fish until the fish die or are caught. And where do these contaminated fish go when they get caught? To our supermarket shelves and ultimately into us. Yes, we and the fish are becoming plastic cybernetic organisms—victims of a contaminated food chain.

Researchers at the University of California Riverside studied graphene oxide, a common form of graphene, to determine how it would degrade when left to nature.[7] What they found was interesting and unexpected. Graphene oxide in open water tended to remain stable, which means the life there would be exposed to it or consume it, much like our super-silk producing spiders (recall that some of the exposed spiders died), and just like fish consuming plastic. Graphene oxide in groundwater, however, tended to break down or settle out, reducing the risk to wildlife.

There have been very few studies on the safety of graphene, and the verdict is definitely out. Not being experts at environment risk assessment, we would like to quote from a 2014 interview with the National Science Foundation's International Chair of Environmental Health Sciences, Dr. Andrew Maynard, that appeared in the July 2014 issue of the *Graphene Council Newsletter*:

> As with any chemical or material, the rules of safe design and use need to be developed if materials like graphene are to be utilized effectively. Increasingly, commercial success will depend on innovating responsibly—taking account of the environmental and societal benefits and impacts of a product as well as its technological and economical viability. This will require relevant research on exposure, hazard and risk. But it will also depend on bounding that research and how it is applied by considering plausible use and exposure scenarios as well as plausible risks. One of the greatest challenges to developing and using new materials is that it is impossible to prove a negative—to show through research that something is completely safe. Because of this, there needs to be reasonable boundaries placed on what is considered safe enough, and what is a reasonable research questions [*sic*]. Without these, there's a danger that relatively safe materials will suck up precious research time and funding, while potentially dangerous new materials slip under the radar of scientific scrutiny.[8]

In other words, we need to be careful, not panic, take sensible precautions to limit the excessive release of graphene into the environment, and conduct some rigorous studies to determine what the actual risk may be. Nothing we humans do has zero impact on the environment. The best we can do is minimize it.

Chapter 11

USING THE REST OF THE TABLE

Is the materials revolution solely focused on materials made from carbon that form interesting and unique geometries? Or will the novel carbon-based materials discussed in this book be merely a few of the many innovations over the next few years? Buckyballs, those soccer-ball-shaped carbon molecules, were hailed as the super material of the century when they were first discovered. While they are still very interesting and useful, they were not the be all end all of materials science research and discovery. Nor were the carbon nanotubes, and neither will be graphene. This is not to say that graphene and its derivatives won't soon rock our world with the many technological innovations that they spawn, but research will continue and it is inevitable that something else will be found that offers yet more technological promise than we can currently imagine. So where are these breakthroughs and how can we find them? Let's first put them in categories to make it easier to understand what is out there and what might be on the horizon.

PROGRAMMABLE MATERIALS

A part of the next materials science revolution may be programmable materials, a subset of matter that can change shape or behavior through the application of an outside signal, whether from an electrical field, application of pressure, or the manipulation of another local property.

You may be getting hands-on working experience with one type of programmable material if you are reading this book on a smartphone or tablet. To navigate to the book app used to find, purchase, or simply open this book, you likely used a touch-sensitive screen. A touchscreen is a transparent material layered over and integrated with your device's visual display and control electronics. When pressure is applied, the screen responds in a preprogrammed way to communicate with the device's electronics to produce the desired result. (Some touchscreens use a change in the electrical properties of the screen when touched to interface and control the device's response. While the physics is very different for pressure versus electrical properties, the underlying functional result is the same.) This can range from turning the device on or off, opening or closing an app, to entering the text that will someday find its way into a book much like this one.

Graphene will likely play a part in making the next-generation touch screen, regardless of the gadget to which it is attached. The graphene component could be as simple as the case around the device (to make it stronger), part of the actual display electronics or sensor, or merely as a lightweight, strong screen protector to help keep it from being damaged during use.

To continue our survey of programmable materials that are in common use today, we will consider Nitinol. Nitinol is an alloy of nickel and titanium that can be shaped into one form and then, when heated, changes shape on its own. It is often made into a wire and can be applied to a variety of consumer and industrial products. Nitinol is used in a variety of applications, ranging from the braces on your children's teeth (where the body heats the Nitinol wire, causing it to contract and apply the force necessary to correct tooth placement), in the stent your grandmother had placed in her heart during surgery, in thermostats (where its shape change is temperature dependent) and to control the shape of systems in space that can't be easily repaired if motors break down. The coolest part? Nitinol was discovered in 1959 and scientists have been finding more uses for it nearly every year since then.

What if we can extend strong shape memory materials into very practical aspects of our daily lives, like repairing damage to our car in a parking lot fender-bender accident? It isn't too difficult to imagine having our car bumpers or side panels made from a material that is designed to assume one shape when heated or exposed to a certain wavelength of light and another shape otherwise. Instead of taking your car in for an expensive repair or installation of a replacement part, technicians might simply expose the bumper to their finely tuned "car repair light," so that it returns to its original shape. The technology might also be applied to aircraft, allowing them to alter their shape depending upon the conditions in which they are flying, optimizing their performance according the local conditions. Most modes of transportation are made to maintain a single static form. What if, instead, the shape of the vehicle could subtly change its shape to recognize local environmental conditions? The car, plane, or boat could slightly shift its outer shell to boost fuel efficiency by a few percentage points, like a professional biker makes minute adjustments to their posture on a downhill slope to eke out every last bit of speed.

Jahn-Teller metals, with environmentally dependent electronic properties, are prime examples of programmable materials. Recent experiments merge programmable materials with one of graphene's cousins, the buckyball. A sixty-carbon buckyball, saturated with the metal rubidium, can, with the application of pressure, be morphed into a football shape and then returned to a spherical shape once the pressure subsides. Responsive molecules such as this one could hold the key to controlling any number of "off/on" systems at the monomolecular level. "Off/on" systems, you may recall, are the basis of the digital revolution, and they would be very valuable additions to a number of technologies.

Other materials are also under investigation for their off/on potential, as well as for some of their other, more exotic, chemical properties. *Catenanes* and *rotaxanes* are two classes of nanomaterials under the umbrella of Mechanically Interlocked Molecular Architectures (MIMAs). You can think of catenanes as molecular magician rings and rotaxanes as molecular dumbbells. Popularized in 1983 and 1991, respectively, catenanes and rotaxanes won Jean-Pierre Sauvage and Sir Fraser Stoddart two-thirds of the 2016 Nobel Prize in Chemistry "for the design and synthesis of molecular machines."[1] Catenanes and rotaxanes had been created forty years prior to Sauvage and Stoddart's contributions, but it was their work that allowed for the efficient production of these new molecular machines. We will cover the third winner later in the chapter when we talk about molecular motors.

Catenanes and rotaxanes aren't perfectly described as molecules, per se. That's because molecules are defined by the electron-sharing between all atoms in the structure. Instead, catenanes and rotaxanes are more correctly described as two separate molecule pieces that happen to overlap with one another. Catenanes are MIMAs that look like two rings locked into one another, as seen on the left in figure 11-1. They are made from long chains of molecules (the exact makeup can vary) that are bent around and closed on themselves to be permanently attached. The rings are still attracted to one another, though through weaker intermolecular forces, like the forces between sheets of graphene. These intermolecular forces create what is called a *supramolecular* system, a system that is more than just one isolated molecule. Rotaxanes, on the other hand, resemble a dumbbell with an unattached ring encircling the handle. The end "weights" are the bulky parts of the molecule that keep the ring from sliding off. The places where the ring and handle strongly interact are called *stations*, and the ring can shuttle or jump between stations when the conditions are right.

How did the ring get onto the rotaxane in the first place, then? How *does* the proverbial ship get into the bottle? Researchers have been able to find, over the course of many years of experiments, that they can specifically predesign an attraction between the ring and handle so they will self-thread. Once this has happens, another chemical reaction will add the weights to the ring/handle supramolecular system, trapping the ring as a part of the whole system!

But what do these weird not-molecular molecules have to do with binary logic (computer) systems? Catenanes and rotaxanes are basically nanomachines; they are the first examples of nanobots in action. These molecules are not self-replicating, therefore this chapter is not a harbinger of the nanobot apocalypse and they won't turn your car into gray goo. Nanomachines will eventually become an integral part of as many different applications as will graphene. It is almost a certainty that graphene will find itself incorporated *alongside* these nanomachines for various applications. As mentioned throughout this book, graphene is certain to be a wonder material—but it will require working in conjunction with other materials to truly shine.

Figure 11-1: Catenanes (left) and rotaxanes (right) are made up of complex molecules that can be simplified in this schematic. (Left) Interlocked rings (silver and black) form from inter-ring attraction. (Right) A ring is threaded around a handle (solid black), which has two stable stations (patterned boxes). The capping dumbbells prevent the ring from slipping off. (Image by Joseph Meany.)

Interest in MIMAs is not a purely academic exercise. Rotaxanes are primed to generate significant interest in unusual situations. This was made clear in 2005 when a collaboration between research groups in the Netherlands, the United Kingdom, and Italy produced nanomachines that could push liquid uphill with just the addition of some light.[2] The groups produced a rotaxane that had two stable stations along the handle,[3] and the handle was attached by one of the weights to a specially prepared surface. Normally a liquid will roll down a surface upon which it is placed. This should not come as any shock to anyone—gravity works. The researchers found that, under normal conditions, a rotaxane-modified surface rolled liquid downhill. However, when the rotaxane-modified surface was illuminated by a special light, the liquid flowed uphill, against the flow of gravity.

What was happening there? Essentially, when the light hit the surface, it was absorbed by the rotaxane ring. This gave the ring enough energy to shuttle from one stop to the other. Jumping to the second stop caused the top weight to repel the liquid. By carefully pointing where the light shined, the researchers were able to make the whole droplet to roll *up* the special wafer. When the light was turned off, the ring jumped back to its original stop, and the liquid rolled back down the wafer. This was an excellent example of forces on the quantum scale adding up to produce a big, "real world" effect.

The arena of nanomachines is not only limited to these two curiosities. To realize true microrobots, machines will have to work in concert with the random noise caused by heat in the local environment. Medicinal microbots would be useless if their pincers or propellers could not function at normal body temperatures. The human body stays at around 37°C (98.6°F), which is about 310 Kelvin. At this temperature, there is a lot of heat energy available to shake and vibrate the atoms. They jostle around one another like powered billiard balls, each rebound sending a molecule or atom off to its next chaotic collision in the quantum quagmire. In this kind of hostile environment, it seems almost impossible to imagine that intentional motion on this scale might be possible.

Bacteria can move with intention. Cells within our bodies also move with intention. We know that biological organisms have devised clever ways to get around this planet and scientists are learning from these organisms; in order for tiny machines to be truly viable, they must be able to move intentionally. In 1999, Bernard L. Feringa and his group created a light-driven molecular-scale motor.[4] This molecular motor was capable of moving only in a single direction, which is a necessary function for anything called a machine. Other materials had come close to working before this, and others since 1999 have improved upon the system Feringa developed. His motor was the first gold-standard, though. For this work, Feringa was the third recipient of the 2016 Nobel Prize in Chemistry, alongside Sauvage and Stoddart.

Almost twenty years have passed since Feringa's first molecular motor and progress continues toward creating molecules that can perform controlled functions reliably. As this book was being written, for example. some of the greatest molecular machinists in the world were fresh off the finish line at the world's first-ever nanocar race.[5] Nanocars are an exciting prospect for nanomaterials research; they are little molecule-sized vehicles that can roll, slide, or glide across a surface to perform whatever necessary function they are designed to accomplish. Applications can include acting as passive environmental sensors or delivering a DNA-tagged ribbon of graphene to some experimental cell. These nanocars could move molecules for purposes that natural random molecular motion would be just far too imprecise to trust.

In 2005, Professor James Tour and his associates made the first nanocar at the Tour lab at Rice University. They created a carbon-based molecule with a rigid chassis to which four fullerene balls were attached, acting as wheels. The group found that adding an electrical charge to the molecule caused it to roll across an atomically flat gold electrode. This demonstration laid the foundation for a nanocar race—which the Tour lab participated in and, unsurprisingly, won.

We need not be bogged down by thinking about microbots or nanocars as being confined only to buzzing about a surface in two dimensions. Additional breakthroughs in materials science are going to occur from advances in three-dimensional movement as well. Once we master controlled motion in six *degrees of freedom*,[6] then we will be able to harness the full potential of atomic control within complicated systems. Richard Feynman talked about microbots in medicine during his 1959 lecture "There's Plenty of Room at the Bottom," which we mentioned in chapter 2.

Not long afterward Feynman's lecture, the shrinking-submarine-in-the-bloodstream became a science fiction staple in the film *Fantastic Voyage*, based on a story by Otto Klement and Jerome Bixby. In *Fantastic Voyage*, scientists are hit with a "shrink ray" and injected into a fellow scientist's blood stream to remove a blood clot from his brain. The science-fiction shrink ray miniaturized the scientists' atoms, which is certainly not possible, but the idea behind it, noninvasive precision surgery from an injected object, is certainly viable. *The Magic School Bus* television program also had an episode based on the same concept, in which Mrs. Frizzle and her students went "Inside Ralphie" to learn about bacterial infections in the bloodstream.

Unfortunately, we cannot rely on magic busses or shrink rays to perform futuristic medical miracles. We must instead rely on factual scientific principles. If a nanobot will run on fuel like a microscopic jet fighter, then it will need to scavenge fuel from whatever environment into which it is placed. If the nanobot is going to be propelled by blood pressure, then it will need to be able to somehow steer its rudder to turn the device toward and through a specific blood vessel. Some researchers are taking inspiration from biological systems and have designed tiny corkscrew motors similar to the flagella on a sperm cell or the bacterium *Helicobacter pylori*. This approach is motivated by the fact that water and other fluids act differently on the microscale level than we are used to in our macroscale world. While you might think the water in a cup is one continuous substance, cells and microbots would encounter water more as a mass of jostling, always-moving, bouncing balls. If you or your child have ever tried to swim through a fast food restaurant's playground ball pit, then you have an inkling of what a microbot would be trying to swim through in your body.

We are easily several decades from seeing microbots widely used in medical procedures, but contemporary advances in body-imaging techniques will allow us to track and guide future microsurgeons. Someday, microbots could be adapted for use in water treatment, grasping onto heavy metals and other toxic compounds in our waste streams and removing them. There is even a possibility that microbots could assist in space travel, becoming our own little versions of Star Wars's R2 units that could patch leaks from micrometeor punctures or solar radiation erosion. We have the capability to find an application for almost any material, putting it to good use in space and here on Earth.

THE POSSIBILITIES OF ADDITIONAL TWO-DIMENSIONAL MATERIALS

Excitement has grown over the past twenty years over new materials forged by the cooperative enterprises of physics, chemistry, and materials science. Computers have allowed researchers to build new physical models to predict structure-property relationships. Recall that nanomaterials can be classified based on their relative dimensions. Quantum dots and fullerenes, because they are highly symmetric and have no dominating direction, are zero-dimensional. Carbon nanotubes have two short dimensions in the *y* and *z* directions; properties are defined by a nanotube's length along the *x* direction and are thus called one-dimensional. Three-dimensional materials, on the other hand, are the types of materials that have no dominating directionality but are large enough to see and hold. These three types of materials were once believed to be the only types possible.

Early experiments on thin films with single-atom thicknesses failed. Evaporating metals to form single-atom sheets produced globular islands instead. It seemed that two-dimensional materials would not be possible—until the isolation of graphene sparked a wildfire of new research. Billions of dollars have been granted for research and development, and the research is not limited to only carbon-based graphene. Other new materials have also been found that extend only in the *x* and *y* directions. These two-dimensional materials are interesting chemically because they open the door between the world of quantum physics and the world of our typical understanding. Graphene is the most obvious example, but other two-dimensional materials are being discovered all the time. The following section gets into some of the more interesting materials, such as the graphene analogs from other elements within carbon's periodic group[7]—the *Xenes*. Other examples will cover the elements to the left and right of carbon, boron and nitrogen, respectively.

Graphene was a remarkable discovery, but we would be remiss to ignore the other emerging two-dimensional materials. These materials will work in conjunction with graphene to create other new and unusual devices. It would not be correct to directly relate all two-dimensional materials to graphene—they will almost certainly not be the same, or even close. Some may be conductors (like graphene), some not. Some will be structurally strong (again, like graphene), others not. You get the idea. Andre Geim was looking for new and interesting questions to answer when he and Konstantin Novoselov isolated graphene in 2004. They found new and interesting questions where they didn't expect them, and other scientists have since taken their seminal work and expanded it into an incredible foray of scientific discovery. The graphene revolution is only one (admittedly exciting) aspect of the greater drive to apply principles from the nanoscale to medicine, clothing, and space travel—all macroscale applications.

Graphene, as we have hopefully made clear, is made from carbon atoms that are connected to one another more strongly than those in diamond. They are connected in a way that allows the *p*-orbitals above and below the carbon plane to mesh together, leading to the superlative electronic properties of graphene. Carbon is flexible. Carbon's ability to form one, two, or three bonds of varying geometries make it even more versatile than we have explained thus far. In the veritable library of carbon-based molecules that we find in nature, from the depths of the Mariana Trench out to the atmospheres of stars, carbon makes shapes ranging from simple equilateral triangles to the globs of carbon in candle flames. It should come as little surprise, then, that the advent of graphene would inspire chemists to dream up and look for other exotic forms.

Graphene, under the right conditions, may be *reduced*; that is, those *p*-orbitals can bind with other atoms instead of with their carbon neighbors. Hydrogen was the most obvious first choice for this kind of reaction, called *reduction*. If you think back to one of the earlier chapters, graphene was named as such because of its relation to graphite. Its delocalized, double-bonded nature earned it the "-ene" suffix. This is common in organic chemistry—for example, C_2H_4, ethylene (or ethene, more properly), can be reduced by adding hydrogen to its structure and making C_2H_6, ethane. It follows that by adding hydrogen to graphene, the two-dimensional crystal would be reduced to graph*ane*—where each carbon atom gets its very own hydrogen atom. The reduction changes the very way that graphene functions in this case, rendering it nonconductive. But why do we care? Why would this be useful at all? I mean, we have spent this whole book talking about how great the conductivity of graphene is, and there are people out there actively trying to ruin that? That just seems so unusual, unproductive even. But it is the process of science to investigate a new material in all its forms.

This is a case where the high surface area of the graphene is its strength. Every carbon is exposed to the surface. When hydrogen is introduced to graphene supported on a surface, the hydrogen will attach to half of the carbon atoms on that one face. Hydrogen does not attach to every carbon because they would crowd each other out. It is an interesting fact of the chemistry between carbon and hydrogen that this bond is not particularly strong. Graphane may find a potential use as a source for rechargeable hydrogen fuel cells, expanding the scope of graphene and its derivatives and contributing to the generation, management, and use of electrical power. When heated to 450°C, the graphane releases the hydrogen atoms and allows the hydrogen to be gathered for energy use. This chemically turns the graphane back into graphene, which, when cooled, can accept more hydrogen, making it into a rechargeable power source. Can you think of something that you put fuel in, store for a while, and then use as needed? Something, perhaps, like a car? Most people understand by now that fossil fuels are unsustainable as an energy resource. One potential replacement for oil and coal in producing energy is hydrogen fuel cells.

One of the problems keeping hydrogen fuel cells from becoming widely used in our cars is the production of hydrogen. Since matter cannot be conjured from nothing, a source of hydrogen is required. State of the art fuel cells get their hydrogen from water or oil. Energy is required to remove the hydrogen atoms from either source, and here you can see the conundrum. You must spend energy to produce energy, and the laws of thermodynamics require losses each step along the way. The next problem is finding a compound or system suitable to carry and hold the hydrogen once it has been produced. Methane (CH_4) and diborane (B_2H_6) were two early candidates for storing hydrogen to be used in fuel cells. Diborane and methane are both gases, though, and there is a concern that storing these gasses at high enough pressures to produce useful amounts of energy is a safety hazard. Anyone who has ever seen a can of butane burst will nod their head in agreement. On top of this, diborane has the troublesome property of being pyrophoric. If it touches air, especially the warm moist air of a summer day, then it will burst into flame explosively. This is a double no-no for fuel cell use. Graphane, as a solid, subverts these problems. It does not require high pressures or special handling under inert atmospheres to safely store. Its stability when charging and discharging may finally open up doors that have eluded other materials. Where hydrides have failed, graphene-derived fuel cells may finally find a home.

We discussed early in the book how the geometry of carbon-carbon bonds created the fullerene and carbon nanotube structures. Five carbon atoms in a ring pucker the structure from a flat plane into three dimensions. Likewise, seven carbon atoms in a ring will force the creation of a three-dimensional structure. With proper introduction of five-member and seven-member rings into a sheet of graphene, three-dimensional structures could (hypothetically speaking) be made, and a carefully designed hollow tubular structure could support not only hydrogen adsorption on one face of the graphene, but on both faces. This would more than double the efficiency of a graphane hydrogen cell, allowing it to be constructed in a familiar block-like shape rather than requiring a large, flat surface to store meaningful quantities of hydrogen.

If we can use chemistry to create three-dimensional shapes out of carbon-based graphene relatives just by changing out the number of atoms in a ring, are there any other ways that carbon can be arranged to create two-dimensional materials? Theoretical chemists say yes. The linear chains called alkynes (much the same as the ones that Harold Kroto was looking for in stellar atmospheres) can be interspersed with the aromatic rings to form a new type of conducing two-dimensional material called graph*yne*. This is the natural extension of graphene and graphane, all named for the -ane → -ene → -yne progression used in all other hydrocarbons. As there are ethane (C_2H_6) and ethene (C_2H_4), there is also ethyne (C_2H_2), better known as acetylene. Common names, which do not follow a standard nomenclature, can be confusing. Graphyne is the simplest example of a possible wealth of new carbon allotropes. Introducing -C≡C- spacers separating the rings of graphene would allow for holes to be "programmed" into the two-dimensional structure. In other words, atomic filters could be manufactured with specifications tailored to removing a contaminant of interest. The size of these pores could be designed for any purpose just by varying the number of the alkyne units bridging the rings. (Recall the water filters mentioned in chapter 6.)

And this brings us back to the excitement behind carbon nanotechnology. We are finally beginning to realize that for any conceivable problem, we can imagine and develop a uniquely suitable material to solve that problem. Carbon is a suitable banner for this cause, as it is a familiar element to the taxpayers and investors who will need to be the funding these opportunities behind commercialization. Research on nanoscale materials will certainly continue toward other elements beyond carbon; as the complexity (and therefore accuracy) of computer models grows over time, researchers are better able to predict how a potential material may behave. Buckminster Fuller thought through this problem, once saying about architectural structures, "The last tensile wires will simply be the chemical bonds."[8]

This high-throughput screening allows theorists to test out molecules and compounds that are not cost-effective to develop physically. At least, not at first. We still have ninety-one-odd other stable elements on the periodic table to work with, and we only have a passing knowledge about most of them. Is there any way that other elements can make graphene-like structures whose properties might be equally miraculous? There is a concept within chemistry that explains how compounds with similar bonding arrangements may behave similarly—*isoelectronics*. To be isoelectronic with graphene, a material would need to have a closely related arrangement of electrons in its orbital clouds. Elements in the same column as carbon (silicon, lead, etc.) are automatically isoelectronic with carbon, which means that we have been able to create and study *-ene* molecules based on graphene-like hexagons of these other elements. Silicon can be arranged into silicene. Germanium gives germanene. Those are straightforward and easy to remember. Unfortunately, tin and lead are different; they aren't called tinnene or leadene. Rather, tin's elemental symbol is Sn for the Latin *stannum*. Thus, a tin-based graphene would be stannene. Likewise, lead's symbol is Pb for the Latin *plumbum*, and so its graphene-equivalent would be plumbene.

We know even less about these isoelectronic analogs to graphene than we do about graphene itself. We had at least an extra hundred years of research to understand graphite before we got to graphene. With silicene or plumbene, we know of no natural mineral containing two-dimensional sheets of silicon or lead.[9] Human ingenuity is not to be outdone. From the time of graphene's isolation onwards, each of the isoelectronic graphene analogs has revealed itself to determined researchers. Plumbene was made in 2004.[10] Silicene was made in 2012.[11] Germanene was made in 2013.[12] Stanene, the last to fall, was confirmed in 2015.[13] Each has provided new lines of inquiry to follow down the rabbit hole of physics as we explore ways we might exploit the laws of nature for our benefit.

Isoelectronic compounds are not merely limited to the column below carbon, though. Combining elements from the columns left of carbon with elements from the columns right of carbon would also produce a hexagon lattice. Hexagonal boron nitride, h-BN, is a graphene-like two-dimensional layer made from boron (left of carbon) and nitrogen (right of carbon). Boron has one fewer electron than carbon; nitrogen has one more. When the two elements react together, they form a hexagonal structure equivalent to that of graphene. Nitrogen has a paired electron orbital, and this orbital donates both electrons into the orbital where boron lacks any. This scenario is electrically analogous to two carbons donating one electron apiece to each other. This planar-hexagon version of boron nitride (there is also a cubic version with a crystal comparable to diamond) is well regarded for its lubrication properties—like graphene, it also readily cleaves along its crystal planes. However, unlike graphene, h-BN is not conductive. In fact, it is such a poor conductor that it would be more appropriate to consider it an insulator. A flat, two-dimensional insulator such as this is an exceptional boon to the nanomaterials community, however. Consider that graphene is a nearly perfect conductor. It takes no input of energy to promote an electron from the valence (tightly held) band of electrons in graphene to the conduction (loose) band of electrons within the crystal. For example, LED lights work by promoting electrons from the valence to the conduction band. A red light, the lowest energy light, only requires about two or three electron volts to work. This is low energy, as far as semiconductors are concerned. A yellow or green light require higher energies—about three or four electron volts. Blue and violet lights are made with band gaps just above four electron volts, and above five electron volts the LEDs emit ultraviolet light. Beyond this, LED lights are no longer useful. H-BN has a bandgap of 5.9 electron volts. This is incredibly high and could only find use as a semiconductor in very specific applications. In more general situations, boron nitride is a very handy material for *preventing* the flow of electricity. Due to the nonexistent bandgap of graphene, the material appears black to our eyes because it can absorb all of the different wavelengths of light that we can see. Boron nitride, on the other hand, absorbs no wavelengths that we can see. Due to its high bandgap, it can reflect all visible wavelengths. Therefore, the mixture of boron nitride's flat hexagonal crystal structure, along with its highly reflective nature, has prompted boron nitride to be nicknamed "white graphene." White graphene is a tough material and is used as a lubricant where graphite lubricants would not be possible. This has been its primary application since the mid-1940s. After the isolation of graphene in 2004, h-BN became an obvious material to pair with graphene to create new electronic devices. In 2010, h-BN was used to sandwich a layer of graphene. The two pieces of boron nitride "bread" isolated the graphene chemically and electronically from all other interactions with the environment, allowing the researchers to test an exceptionally "clean" system of materials without defects present. The boron nitride insulated the conductor in the same way that rubber insulates wires in your house. This sandwich approach set a record for graphene conduction. While world records are nice, this finding also confirmed a deeper truth about graphene—it is still moderately reactive with the environment, even if just barely so, and that reactivity affects its properties. It suggests that if we want devices that will utilize graphene to its highest potential, then we will still need to protect it from outside interference with something like pristine h-BN. A "real world" high-current conducting cable will probably need to be a long, unbroken carbon nanotube coated in a sheath of h-BN nanotube to protect it from weathering. Nothing, it seems, is simple!

Occasionally during the synthesis of h-BN, graphene may be introduced into the mix, as well, which produces a complex material called a borocarbonitride (BCN for short). The exact properties of a BCN are going to vary widely depending on many different factors, especially the conditions under which it is manufactured, making it difficult to talk about what properties *all* BCNs would share.

While h-BN and BCN are examples made from a metalloid (boron) combined with nonmetals (carbon, nitrogen), other special two-dimensional materials can be made from metals combined with nonmetals. This general class of materials are called MXenes.[14] The M stands in for some transition metal, which is one of the elements toward the center of the periodic table. The X is a placeholder for other nonmetals, listed on the right-hand side of the periodic table. When combined, the metals and nonmetals can form an extended two-dimensional crystal. These crystals aren't usually flat. More often than not, the crystal is buckled in some way. Note that the large lateral area compared to the thickness is what drives the "2-D material" designation for this class of compounds. As large, laterally spread crystalline materials have grown in number, researchers have added them to the list of two-dimensional materials, whether they are atomically thin or not. This seemingly blurs the line of what a two-dimensional material *is*, but it is important to remember that the dominant physical behavior is the defining factor for whether a material is two-dimensional (or not). If propagation of some signal—whether that signal is a photon, an electron, or a vibration—is significantly diminished in one axis against the other two[15] then it will be considered a two-dimensional material.

With the rapidity with which materials are being created and tested, this chapter cannot properly cover *all* two-dimensional materials. Science was given an incredible gift in graphene, and the wealth of derivative materials that have been discovered since will continue to grow by leaps and bounds. Compounding the complexity of summarizing two-dimensional materials would be the impossible task of summarizing other materials of zero, one, and three dimensions. The example of h-BN nanotubes provides a hopeful glimpse that where carbon fails an application, some other molecular combination may yet prevail. MXene buckyballs may suddenly appear and show that you *can* reversibly open and close a cage to deliver payloads in a body. Or maybe BCN nanotubes will become the microtractors on a nanofarm growing custom proteins. The frenzy of activity in nanoscience kicked off by the graphene revolution will lead to further research with the rest of the periodic table. Chemistry in one hundred years will likely look upon our knowledge today in the same way we view the alchemists from three hundred years ago.

A POSSIBLE FUTURE

Futurists, who are often engineers or scientists, predict a near-future in which everything in our lives is multipurposed, thanks, at least in part, to shape changing or programmable materials and materials with properties beyond the current state of the art. Consider the home of the future that can transform a wall into a door upon command or, more simply, darken otherwise transparent windows using perhaps an electric field or current to alter the reflectivity or absorption of a material coating it. What if that new outfit you purchased at the store or online could be made to change color or style by simply asking it to do so? (According, of course, to a set of preprogrammed alternatives somehow stored in the material or via the assumed-to-be ubiquitous cloud.)

And what if everything is like this? What if our entire world can be repurposed, remade, or remanufactured to meet our material demands without the need of multiple, potentially redundant products lining our garages? Combine the functionality of programmable materials with the amazing mechanical, electrical, and structural properties of graphene and similar materials, and it is possible we will soon experience the end of our throwaway culture. Along with the end to this culture would also mean the end of most pollution, which will only be a benefit to our current planet and any others we may seek to inhabit. And it all might be made possible by one of the most abundant, most versatile, and most essential of all elements, carbon. The same carbon that forms the basis of all known forms of life on Earth and that enables graphene to be formed. Graphene—the superstrong, superthin, and superversatile material that will revolutionize the world.

AFTERWORD

What is next for graphene? How will this potentially revolutionary material transition from university laboratories to the market and then on to changing the world? The answer is not easy. There are those who reside in the "build it and they will come" category. These folks have an unshakable belief in the marketplace of ideas and market-driven economics. If the product is superior to the competition and its price is competitive (or low), then people will buy it and make the technology successful. There is a great deal of truth in this belief, and, at least on the surface, history seems to support this viewpoint; good new ideas can often become successful in the marketplace for exactly these reasons. The smartphone is a great example. No one was asking for an iPhone, but once people saw the capabilities offered by one they could not imagine doing without it. Smartphones are now considered one of the most successful product innovations in history, and they were quickly adopted around the developed world.

Even if "build it and they will come" is the correct analog for the development of graphene-based products, then there is still the problem of "building it." A cheap, efficient, and high-volume mass production system has not been successful to date. Likewise, early entrants into the graphene market have resisted efforts to standardize easy and cheap analysis methods, so there is not yet a reliable source for the raw material that might be used in graphene-enabled or enhanced products. Fortunately, various companies, nongovernmental organizations, and research institutions are certainly working to make that happen. When it does eventually occur, customers will finally be able to most accurately assess what product they will need. In this approach, there will likely be multiple suppliers competing for sales, each offering slightly different types of graphene and of varying quality.

Rather than rely on seemingly-random market forces, some countries are making conscious efforts to foster graphene-related research through financial resources devoted to development and innovation. This is being done through grants and contracts, tax incentives, and various other methods that governments have at their disposal for fostering innovation. Most notable in this category is the National Graphene Institute in Manchester, England. The institute, funded with over £38 million from the UK Government and £23 million from the European Regional Development Fund, is a partnership of over forty companies working with researchers from the University of Manchester toward making the graphene revolution happen—with UK companies as the primary beneficiaries. Their soon-to-open Graphene Engineering Innovation Centre will increase their overall research capabilities and bring in even more collaborators. The University of Manchester appears to be the focus of graphene research within Europe.

In 2013, China established their own institute, the China Innovation Alliance of the Graphene Industry (CGIA). CGIA, like most Chinese research consortia, is not as well-known as its European or American counterparts, but it is nonetheless a graphene research, development, and commercialization powerhouse.

What about the United States? For the most part, graphene research and development in the United States is decentralized and only loosely coordinated by the various government laboratories, universities, and commercial companies performing graphene research. To make the coordination of American efforts easier, the National Graphene Association (NGA) was formed in 2017. NGA's goal is to help American innovators get their graphene-related products to market as quickly as possible. An admirable goal. And one that is aligned with similar institutes and consortia in other countries around the world.

It is said about remarkable scientific breakthroughs that, through the lens of history, it seems almost as if pure and mature ideas spring forth fully formed from the minds of their creators. Democritus, Boyle, Newton, Curie, Einstein, and Bohr are all popularly acclaimed to have had these flashes of insight. However, we do them a great injustice to reduce their conclusions to what otherwise could be mistaken for divine revelation. Passion, curiosity, and a relentless desire to find order in our natural world are the true gifts that these great minds possessed. These gifts they passed onto us, that we may follow in their footsteps to appreciate the full beauty and splendor that is our universe. We see this gift from Dresselhaus, Geim, Novoselov, Acheson, Humphry, and all the others mentioned in this book. Many others, especially the current graduate students and postdocs at the benches, will provide further incremental understanding and, perhaps, the next flashes of insight and innovation. Their work are the steps up a mountain, leading fellow climbers to a summit and a new horizon. More beautiful is the fact that just about everyone has this same capability to put in the work, the capacity to ask questions, and the capacity to create knowledge from their ideas. And then, in the course of history, another curious mind will pick up the trail where the first creator left off. Anyone can become a giant upon whose shoulders another can stand; we truly stand on the shoulders of our forebears and are creating new giants every day.

As you have read throughout this book, the history of carbon science as a whole, and even graphene in particular, has benefitted from the input of diverse ideas transcending all ideological boundaries. Graphene and other two-dimensional materials stand poised at a junction to generate a proliferation of special compounds touching all aspects of our lives. The research has benefitted from individuals collaborating across oceans, aided by the internet. It has benefitted from journalists touting its extreme behavior in often sensationalist articles. It will continue to benefit from sustained research in both nonprofit and for-profit sectors. Both funding structures will be required to bring our supermaterial to its highest potential. Development of a mass-production mechanism will not come as an inspiration from on high. Market realization will not be divinely gifted. These things will, however, come through a workforce that is free to explore beyond the current boundaries of scientific knowledge. They will come from educated individuals who can focus their attention on understanding the past to help create a better future. They will come from the efficient exchange of ideas on many different platforms—perhaps in ways that we do not yet imagine.

A great leap of imagination is not required to see that graphene is poised to change our society in a way that rivals the development of metal tools that took us from the Stone Age to the Bronze Age. We are at the very beginning of the *Graphene Age*.

ACKNOWLEDGMENTS

The authors would like to thank our agent, Laura Wood of FinePrint Literary Management, for giving us the opportunity to write this book. Her text asking, "What do you know about graphene?" began a two-year-long odyssey to the work that is in your hands. We value the editors at Prometheus (Sheila Stewart and Hanna Etu) for the many suggested improvements they recommended—readers everywhere should appreciate good editors! We would also like to thank Dr. Robert Hampson, aka "Speaker to Lab Animals," for his help with understanding the brain science of chapter 10. We would like to also thank the Atlanta-Fulton Public Library System for their assistance in obtaining research materials. Finally, deep appreciation must be expressed of Mother Nature, for giving us this wonderful universe to explore—and write about!

APPENDIX

Overuse of facts, figures, and statistics can make a reader's eyes glaze over, yet some are interested in such things to provide the context in which they view the subject under discussion. In this case, the subject is graphene, its uses, and how soon it will become more a part of our daily lives. To this end, we provide here worldwide statistics, facts, and figures associated with graphene research and product development.

Let's gain some insight into the rapid growth of patents related to graphene, who is patenting, and how the world views the technology using various metrics. According to the United Kingdom's Intellectual Property Office, in their report titled *Graphene: The Worldwide Patent Landscape in 2015*, the total number of global graphene-related patents has grown each year, and exploded in recent years (figure A-1). Data is not yet available for 2015 and more recent years.

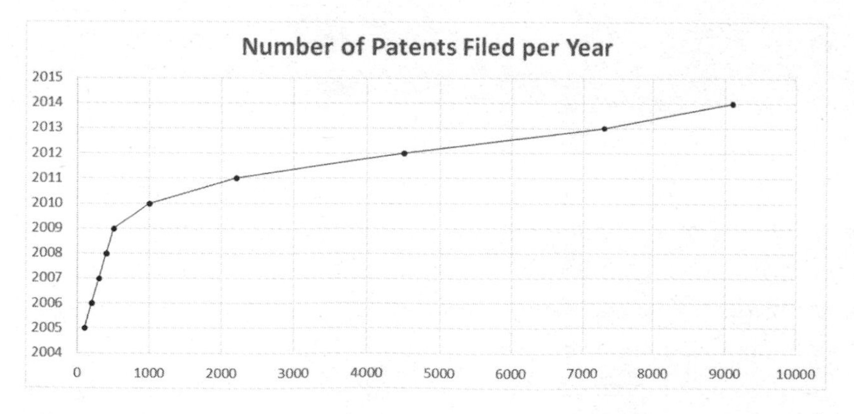

Figure A-1: The number of patent applications filed globally each year since 2004. (Data courtesy of the United Kingdom Intellectual Property Office.)

The data gets very interesting when you learn more about who is filing patent applications. As you might expect, those countries with a healthy academic and industrial research and development base are key players. What might be unexpected is the dominance of China in the data, as seen in figure A-2. In the decade leading up to 2014, the latest years for which the complete dataset is available, China filed for nearly half of the worldwide patent applications. It is important to note that changing tax and patent laws may make such figures a bit misleading. Some inventors may choose to file their patent application in a country other than the one in which they reside for more favorable tax treatment or better patent protections.

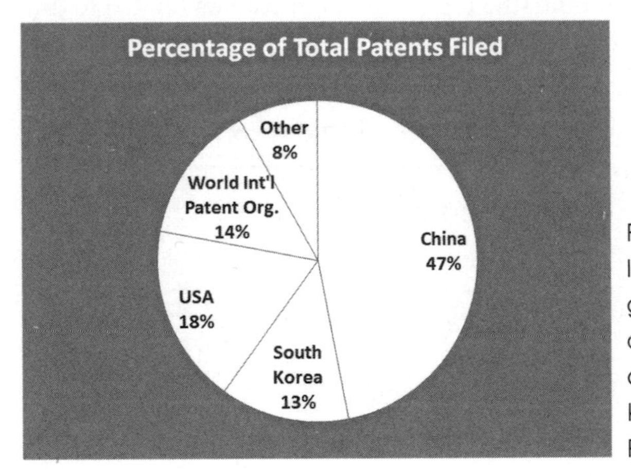

Figure A-2: China is leading the race for graphene patent filing applications. (Data courtesy of the United Kingdom Intellectual Property Office.)

Now let's dig further into the data and determine which universities and companies are responsible for filing these patents (figure A-3). It is here that one can get an idea of where the commercial applications of graphene may soon be coming into play, or least which organizations are funding graphene research. The chart shows "patent families," which are defined as one or more published patents originating from a single original application. There may be multiple patents related to one original patent, each showing some marginal or significant change or improvement, making them all part of a single family.

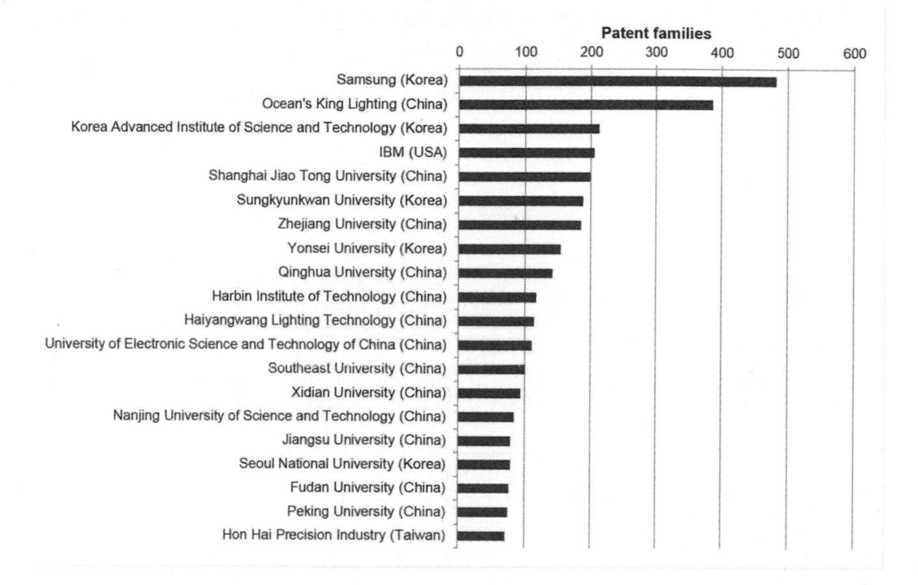

Figure A-3: Based on the number of patent family filings, South Korean companies are clearly leading the way, with China a close second. (Image courtesy of the United Kingdom Intellectual Property Office.)

From the data, it is clear that Samsung is keenly interested in the economic potential of graphene and is funding a considerable amount of graphene-related research. One might wonder where the American technology companies are on this list. IBM is number four on the list and the only American company to make the top twenty.

NOTES

CHAPTER 1: CARBON, CARBON, EVERYWHERE!

1. Okay, maybe not if you're listening to an audio book. You've got us there.

2. Shankar Vallabhajosula, "Science of Atomism: A Brief History," in *Molecular Imaging: Radiopharmaceuticals for PET and SPECT* (Berlin: Springer-Verlag Berlin Heidelberg, 2009), pp. 11–23.

3. Traditional name: Abū ʿAlī al-Ḥusayn ibn ʿAbd Allāh ibn Al-Hasan ibn Ali ibn Sīnā.

4. Traditional name: ʾAbū l-Walīd Muḥammad Ibn ʾAḥmad Ibn Rushd.

5. Georgios S. Limouris, "From the *Atomon* of Democritus to the Therapeutic Nuclear Medicine of Today," *European Journal of Nuclear Medicine and Molecular Imaging* 33, no. S2 (September 2006): S65–68.

6. J. R. Partington, *A Short History of Chemistry* (London: Macmillan, 1957).

7. Vallabhajosula, "Science of Atomism."

8. Limouris, "From the *Atomon* of Democritus."

9. Thomas Thomson, "Of Alchymy," in *The History of Chemistry*, vol. 1 (London: Henry Colburn and Richard Bentley, 1830), pp. 28–29.

10. Ibid.

11. Ibid.

12. Partington, *Short History of Chemistry*.

13. Vallabhajosula, "Science of Atomism."

14. Partington, *Short History of Chemistry*.

15. Francis A. Carey, "Introduction," in *Organic Chemistry* (New York: McGraw Hill, 2006), p. 3.

16. "Justus von Liebig and Friedrich Wöhler," Chemical Heritage Foundation, last updated August 5, 2015, https://www.chemheritage.org/historical-profile/justus-von-liebig-and-friedrich-w%C3%B6hler (accessed May 2017).

17. O. J. Walker, "August Kekulé and the Benzene Problem," *Annals of Science* 4, no. 1 (1939): 34–46.

18. Ibid.

19. Hiroshi Fujihisa et al., "O_8 Cluster Structure of the Epsilon Phase of Solid Oxygen," *Physical Review Letters* 97, no. 8 (August 25, 2006).

20. Humphry Davy, "Some Experiments on the Combustion of the Diamond and Other Carbonaceous Substances," *Philosophical Transactions of the Royal Society of London* 104 (1814): 557–70.

21. *Nobel Lectures, Physics 1901–1921* (Amsterdam: Elsevier, 1967), cited in "Max von Laue—Biographical," NobelPrize.org, https://www.nobelprize.org/nobel_prizes/physics/laureates/1914/laue-bio.html (accessed January 2017).

22. "The Nobel Prize in Physics 1915," NobelPrize.org, https://www.nobelprize.org/nobel_prizes/physics/laureates/1915 (accessed January 2017).

23. "Most Influential British Women in Science," The Royal Society, March 21, 2010, https://royalsociety.org/news/2010/influential-british-women/ (accessed December 2016).

CHAPTER 2: WHAT HAPPENED TO THE OTHER CARBON MIRACLE MATERIALS?

1. Jin Zang et al., "Carbon Science in 2016: Status, Challenges, and Perspectives," *Carbon* 98 (2016): 708–732.

2. Mildred S. Dresselhaus, "Mildred Dresselhaus Bio," The Kavli Prize, May 31, 2012, http://www.kavliprize.org/sites/default/files/%25nid%25/autobiographies_attachments/Mildred_Dresselhaus_Biography_0.pdf (accessed December 2016).

3. Rosalyn Yalow, "Rosalyn Yalow: Biographical," Nobel Media AB, 1977, https://www.nobelprize.org/nobel_prizes/medicine/laureates/1977/yalow-bio.html (accessed May 2017).

4. Research projects are (usually) funded by government grants, with the funding allocated by a committee. The committees are made up of other scientists, and, from this, certain research areas can expand or shrink according to a type of "fashion," and so researchers need to keep abreast of these changes. Their research livelihood is strongly tied into this process.

5. Pencil lead is so called because the first people who dug it up honestly thought it was lead.

6. Alice Dragoon, "The 'What If?' Whiz," *MIT Technology Review*, April 23, 2013, https://www.technologyreview.com/s/513491/the-what-if-whiz/ (accessed December 2016).

7. Yoji Koike et al., "Superconductivity in the Graphite-Potassium Intercalation Compound C_8K," *Journal of Physics and Chemistry of Solids* 41, no. 10 (1980): 1111–18.

8. Mildred S. Dresselhaus, "Future Directions in Carbon Science," *Annual Review of Materials Science* 27 (1997): 1–34.

9. Dresselhaus, "Mildred Dresselhaus Bio."

10. Aka limewater.

11. Calcium carbonate is found as a natural product in many marine environments. Pearls, coral, snail shells—all of these hardened substances are made from calcium carbonate as a building material in the oceans.

12. Phlogiston (*FLOJ-is-ton)* was thought to be a substance in all combustible objects. It is supposedly what made fire burn. Eventually, the reaction between oxygen and combustible materials (coal, hydrogen, even iron) came to be recognized as *oxidation.*

13. John Tyndall, "The Electric Light," *Fragments of Science: A Series of Detached Essays, Addresses, and Reviews* (New York: D. Appleton, 1892), pp. 419–52.

14. Ibid.

15. Catherine M. C. Haines, "Ayrton, Phoebe Sarah (Hertha) nee Marks," in *International Women in Science: A Biographical Dictionary to 1950* (Santa Barbara: ABC-CLIO, 2001), pp. 12–13.

16. This is not the same person as Doctor Mirabilis, Roger Bacon from the early 1200s. That would be ridiculous, some Nicholas Flamel-level elixir of life.

17. "High Performance Carbon Fibers," National Historic Chemical Landmarks, American Chemical Society, September 27, 2003, https://www.acs.org/content/acs/en/education/whatischemistry/landmarks/carbonfibers.html (accessed March 2017).

18. Roger Bacon, "Growth, Structure, and Properties of Graphite Whiskers," *Journal of Applied Physics* 31, no. 2 (1960): 283–90.

19. Ibid.

20. Marc Monthioux, "Who Should Be Given the Credit for the Discovery of Carbon Nanotubes?" *Carbon* 44, no. 9(2006): 1621–23.

21. H. P. Boehm, "The First Observation of Carbon Nanotubes," *Carbon* 35, no. 4 (1997): 581–84.

22. Monthioux, "Who Should Be Given the Credit."

23. Sumio Iijima, "Helical Microtubules of Graphitic Carbon," *Nature* 354, no. 6348 (November 7, 1991): 56–58.

24. Wolfgang Krätschmer et al., "Solid C_{60}: A New Form of Carbon," *Nature* 347, no. 6291 (September 27, 1990): 354–58.

25. "The Nobel Prize in Physics 1956," Nobelprize.org, 2014, http://www.nobelprize.org/nobel_prizes/physics/laureates/1956/ (accessed February 2017).

26. Hyungsub Choi and Cyrus C. M. Mody, "The Long History of Molecular Electronics: Microelectronics Origins of Nanotechnology," *Social Studies of Science* 39, no. 1 (February 2009): 11–50.

27. Richard Feynman, "There's Plenty of Room at the Bottom," (talk originally given on December 29, 1959 at the annual meeting of the American Physical Society) *Caltech Engineering and Science* 23, no. 5 (February 1960): 22–36.

28. "The Nobel Prize in Physics 1965," Nobelprize.org, 2014, http://www.nobelprize.org/nobel_prizes/physics/laureates/1965/ (accessed March 2017).

29. "Clifford G. Shull: Facts," Nobelprize.org, 2014, https://www.nobelprize.org/nobel_prizes/physics/laureates/1994/shull-facts.html (accessed May 2017).

30. M. Mitchell Waldrop, "The Chips Are Down for Moore's Law," *Nature* 530, no. 7589 (February 11, 2016): 144–47.

31. Choi and Mody, "Long History of Molecular Electronics."

32. The whole molecular formula was C_{60}, not $C_{60}H_{12}$ or anything else. The researchers added nitrogen and hydrogen gas into the machine to create various –C–H terminated polyacetylenes and the –C≡N terminated polyacetylenes. To find pure carbon mixed in with all the other molecules was extremely unexpected.

33. If you add weird shapes, you're cheating!

34. "Perfect Shapes in Higher Dimensions—Numberphile," YouTube video, 26:18, posted by "Numberphile," March 23, 2016, https://www.youtube.com/watch?v=2s4TqVAbfz4 (accessed January 2017).

CHAPTER 3: THE DISCOVERY OF GRAPHENE

1. J. Nicastro, in personal communication with Joseph Meany, February 9, 2017.

2. Andre Geim, "Random Walk to Graphene" (lecture; Manchester, UK: University of Manchester, December 8, 2010).

3. Dan Charles, "Ig Nobel to Nobel: Creative (and Fun) Science Wins," *All Things Considered*, NPR, October 5, 2010.

4. Tesla is the name of the unit for *magnetic flux*, named for the famous Serbian-American Nikola Tesla.

5. "HFML Sets World Record with a New 37.5 Tesla Magnet,"

Radboud University, March 31, 2014, http://www.ru.nl/hfml/news/news/news-items/hfml-sets-world/ (accessed March 2013).

6. M. V. Berry and Andre K. Geim, "Of Flying Frogs and Levitrons," *European Journal of Physics* 18, no. 4 (1997): 307–13.

7. "About The Ig® Nobel Prizes," Improbable Research, http://www
.improbable.com/ig/ (accessed March 2017).

8. Transistors are the pieces within computers that handle logic and computation. They are what let programs work.

9. Geim, "Random Walk to Graphene."

10. Remember that these microscopes are able to magnify surfaces so much that a user can see single atoms packed together. It is really quite the powerhouse.

11. Geim, "Random Walk to Graphene."

12. Have you ever gotten a long piece of tape in your hair? It isn't pleasant.

13. Andre Geim, "The Rise of Graphene," *Nature Materials* 6 (2007): 183–91.

14. "The Nobel Prize in Physics 2010," Nobel Media AB, 2010, https://www.nobelprize.org/nobel_prizes/physics/laureates/2010/ (accessed September 2016).

15. Geim, "Random Walk to Graphene."

16. K. S. Novoselov et al., "Electric Field Effect in Atomically Thin Carbon Films," *Science* 306, no. 5696 (2004): 666–69.

17. K. S. Novoselov et al., "Two-Dimensional Atomic Crystals," *Proceedings of the National Academy of Sciences* 102, no. 30 (2005): 10451–453.

18. K. S. Novoselov et al., "Two-Dimensional Gas of Massless Dirac Fermions in Graphene," *Nature* 438 (2005): 197–200.

19. Geim, "Random Walk to Graphene."

20. "Professor Konstantin Novoselov Interviewed about Graphene," YouTube video, 12:10, posted by University of Manchester, March 16,

2012, https://www.youtube.com/watch?v=e8TrTWdzon4 (accessed September 15, 2017).

21. Humphry Davy, "Six Discoveries Delivered Before the Royal Society, at Their Annual Meetings, on the Award of the Royal and the Copley Medals, Preceded by an Address to the Society on the Progress and Prospects of Science," *Edinburgh Review or Critical Journal for June. October 1827* (1827): 352–67.

22. Walt de Heer et al., "Large Area and Structured Epitaxial Graphene Produced by Confinement Controlled Sublimation of Silicon Carbide," *Proceedings of the National Academy of Sciences of the United States of America,* 108, no. 41 (2011): 16900–905.

23. Ibid.

24. Hans-Peter Boehm et al. "Das Adsorptionsverhalten sehr dünner Kohlenstoff-Folien," Zeitschrift für Anorganische und Allgemeine Chemie, 316, no. 3–4 (1962): 119–27.

25. Claire Berger et al., "Ultrathin Epitaxial Graphite: 2D Electron Gas Properties and a Route toward Graphene-Based Nanoelectronics," *Journal of Physical Chemistry B* 52, no. 108 (2004): 19912–16.

CHAPTER 4: A MIRACLE MATERIAL WAITING TO BURST FORTH

1. Zhao Qin et al., "The Mechanics and Design of a Lightweight Three-Dimensional Graphene Assembly," *Science Advances* 3, no. 1 (2017).

2. Graphene Lab Inc., 2017, http://www.graphene3dlab.com/s/home.asp (accessed September 15, 2017).

3. Prachi Patel, "How to Make Graphene: A Simple Way to Deposit Thin Films of Carbon Could Lead to Cheaper Solar Cells," *MIT Technology Review*, April 14, 2008, https://www.technologyreview.com/s/409900/how-to-make-graphene/ (accessed September 15, 2017).

4. Don't you love the way authors say, "Simply heat this or that to

1000 degrees"? As if you can do this on your kitchen stove, which we wouldn't recommend even if it were possible.

5. D. A. Boyd et al., "Single-Step Deposition of High-Mobility Graphene at Reduced Temperatures," *Nature Communications* 6, no. 6620 (2015).

6. Christopher Sorensen, Arjun Nepal, and Gajendra Prasad Singh, "Process for High-Yield Production of Graphene via Detonation of Carbon-Containing Material," US Patent 9,440,857, filed May 10, 2013, and issued September 13, 2016.

7. Did we mention that this process is initiated by a single spark? No high temperature cooking, no complex set of chemical processes, just gas it up and BOOM!

8. Dong Han Seo et al., "Single-Step Ambient-Air Synthesis of Graphene from Renewable Precursors as Electrochemical Genosensor," *Nature Communications* 8, no. 14217 (2017).

9. Jacob Aron, "Make Graphene in Your Kitchen with Soap and a Blender," *New Scientist*, April 20, 2014, https://www.newscientist.com/article/dn25442-make-graphene-in-your-kitchen-with-soap-and-a-blender/ (accessed September 15, 2017).

10. Graphena, www.graphenea.com (accessed September 15, 2017).

11. "Dans les champs de l'observation, le hasard ne favorise que les esprits préparés." Louis Pasteur, "Prononcé a Douai, le 7 Décembre 1854, a L'Occasion de L'Installation Solenelle de la Faculté des Lettres de Douai et de la Faculté de Sciences de Lille"(lecture, Université de Lille, Lille, France, December 7, 1854).

12. Charles Maberry, "On Carborundum," *Journal of the American Chemical Society,* 22 (1900); 706–707.

13. Otto Mühlhaeuser, "On Carborundum," *Journal of the American Chemical Society*, 15 (1893), 411–14.

14. Peter Sutter et al., "Epitaxial Graphene: How Silicon Leaves the Scene," *Nature Materials* 8 (2009): 171–72.

15. Mattias Kruskopf et al., "Comeback of Epitaxial Graphene for

Electronics: Large Area Growth of Bilayer-Free Graphene on SiC," *2D Materials* 3, no. 4 (2016).

16. A. Al-Temimy et al., "Low Temperature Growth of Epitaxial Graphene on SiC Induced by Carbon Evaporation," *Applied Physics Letters* 95, no. 23 (2009).

17. Hint: not a good one.

18. In all likelihood, they would be more highly regarded. Haber was a captain of the German Chemistry Section in the Ministry of War during World War I. He basically led the development of their chemical weapons. Sadly, his work on pesticides after the war resulted in the insecticide Zyklon A, a precursor to the infamous Zyklon B used in World War II gas chambers.

19. Thermodynamics is a cruel mistress.

20. PMMA is a plastic, much like polyacrylonitrile (PAN), which is a common starting material for carbon fibers.

21. Gedeng Ruan et al., "Growth of Graphene from Food, Insects, and Waste," *ACS Nano*. 5, no. 9 (2011): 7601–607.

22. Glowing hot items emit light according to the rules for what are called black bodies, or the black body spectrum. An object's temperature dictates its emitted color, according to the increasing temperature scale of red → orange → yellow → white → blue. "Red-hot" glowing iron has enough heat energy that the vibrating atoms of metal emit light visible to our eyes. Increasing the temperature even more causes the iron to get "white-hot" to our eyes. Blue supergiant stars, some of the brightest cosmic objects, are "blue-hot."

23. No electrons can exist within this energy gap due to the rules governing the quantization of electron states. It is beyond the scope of this book to get into further details, but you can think of the valence-conduction gap as a "no-fly zone" for electrons. They can hop directly from one to the other without traveling the intervening energy state.

24. Lingling Ou et al., "Toxicity of Graphene-Family Nanoparticles: A General Review of the Origins and Mechanisms," *Particle and Fibre Toxicology*. 13, no. 57 (2016).

25. Ibid.

26. Gregg P. Kotchey et al., "Peroxidase-Mediated Biodegradation of Carbon Nanotubes in Vitro and in Vivo," *Advanced Drug Delivery Reviews* 65, no. 15 (2013): 1921–32.

27. Ibid.

28. Ibid.

29. Leon Newman, Kostas Kostarelos, Cyrill Bussy, and Sarah Haigh, "Biodegradation of Graphene and Related Materials in Tissues in Vivo," Graphene NOWNANO, University of Manchester, http://www.graphene-nownano.manchester.ac.uk/our-research/examples-of-current-projects/appl-medi/biodegradation-of-graphene-and-related-materials-in-tissues-in-vivo/ (accessed November 5, 2017).

CHAPTER 5: COMING SOON TO A STORE NEAR YOU? *OR*, SO WHAT?

1. Changgu Lee et al., "Measurement of the Elastic Properties and Intrinsic Strength of Monolayer Graphene," *Science* 321, no. 5887 (July 18, 2008): 385–88.

2. Belle Dumé, "Graphene Has Record-Breaking Strength," *Physics World*, July 17, 2008, http://physicsworld.com/cws/article/news/2008/jul/17/graphene-has-record-breaking-strength (accessed November 1, 2017).

3. Dexter Johnson, "Graphene Heating System Dramatically Reduces Home Energy Costs," *Nanoclast* (blog), *IEEE Spectrum*, June 2, 2015, http://spectrum.ieee.org/nanoclast/green-tech/conservation/graphene-heating-system-dramatically-reduces-home-energy-costs (accessed May 5, 2017).

4. Liang Jie Wong et al., "Towards Graphene Plasmon-Based Free-Electron Infrared to X-Ray Sources," *Nature Photonics* 10, no. 1 (January 2016): 46–52.

5. Jay Bennett, "Graphene-Laced Bike Tires Are Both Stiffer and

Softer," *Popular Mechanics*, March 9, 2016, http://www.popular mechanics.com/adventure/sports/how-to/a19851/graphene-bike-tires/ (accessed April 1, 2017).

6. Yan Huang et al., "From Industrially Weavable and Knittable Highly Conductive Yarns to Large Wearable Energy Storage Textiles," *ACS Nano* 9 no. 5 (May 26, 2015): 4766–75.

7. William McDonough and Michael Braungart, *Cradle to Cradle: Remaking the Way We Make Things* (New York: North Point, 2002), p. 165.

CHAPTER 6: GRAPHENE SUPERCHARGED

1. Carlos I. Calle and Richard B. Kaner, *Graphene-Based Ultra-Light Batteries for Aircraft* (Cocoa Beach, FL: NASA Aeronautics Research Mission Directorate [ARMD], 2014 Seedling Technical Seminar, February 19, 2014).

2. Jens Christian Johannsen et al., "Tunable Carrier Multiplication and Cooling in Graphene," *Nano Letters* 15, no. 1 (2015): 326–31.

3. Qunwei Tang et al., "A Solar Cell That Is Triggered by Sun and Rain," *Angewandte Chemie* 55, no. 17 (April 18, 2016): 5243–46.

4. Zihan Xu et al., "Self-Charged Graphene Battery Harvests Electricity from Thermal Energy of the Environment," *arXiv*: 1203.0161, March 1, 2012.

5. The best way to think of doping is that it is the addition of an impurity. When certain impurities are added, the electrical properties of a substance can be altered, making a nonconductor more conducting or a conductor more insulating.

6. Si Young Lee et al., "Chemically Modulated Band Gap in Bilayer Graphene Memory Transistors with High On/Off Ratio," *ACS Nano* 9, no. 9 (2015): 9034-42.

7. Hyunseob Lim et al., "Structurally Driven One-Dimensional

Electron Confinement in Sub-5-nm Graphene Nanowrinkles," *Nature Communications* 6 (October 23, 2015).

8. J. Hicks et al., "A Wide-Bandgap Metal–Semiconductor–Metal Nanostructure Made Entirely from Graphene," *Nature Physics* 9 (2013): 49–54.

CHAPTER 7: DISRUPTION

1. Arthur L. Schawlow and Charles H. Townes, "Masers and Maser Communications System," US Patent 2929922, filed July 30, 1958, and issued March 22, 1960.

2. Scott McCartney, *ENIAC: The Triumphs and Tragedies of the World's First Computer* (New York: Walker, 1999), p. 5.

3. Peter J. Lee, *Engineering Superconductivity* (New York: Wiley-Interscience, 2001), p. 1.

4. G. Bednorz and K. A. Müller, "Possible HighTc Superconductivity in the Ba−La−Cu−O System," *Zeitschrift für Physik B Condensed Matter* 64, no. 2 (June 1986): 189–93.

5. High temperature being relative only to the very low temperatures required for the superconductors first discovered.

6. Malcolm W. Browne, "Physicists Debunk Claim of a New Kind of Fusion," *New York Times*, May 3, 1989, https://partners.nytimes.com/library/national/science/050399sci-cold-fusion.html?mcubz=0 (accessed August, 28, 2017).

7. *The Graduate*, directed by Mike Nichols (Los Angeles, CA: Embassy Pictures, 1967).

8. Daniel Crespy, Marianne Bozonnet, and Martin Meier, "100 Years of Bakelite, the Material of a 1000 Uses," *Angewandte Chemie International Edition* 47, no. 18 (April 21, 2008): 3322–28.

9. "*Plastics—The Facts 2016: An Analysis of European Plastics Production, Demand, and Waste Data*," (Brussels: Plastics Europe, 2016),

http://www.plasticseurope.org/documents/document/20161014113313-plastics_the_facts_2016_final_version.pdf (accessed August 28, 2017).

10. Gaelle Gourmelon, "Global Plastic Production Rises, Recycling Lags," *Vital Signs*, January 27, 2015, Worldwatch Institute, http://vitalsigns.worldwatch.org/sites/default/files/vital_signs_trend_plastic_full_pdf.pdf (accessed November 9, 2017).

11. *Cast Away*, directed by Robert Zemeckis (Los Angeles: 20th Century Fox, 2000).

CHAPTER 8: OBSTACLES

1. *Wikipedia*, s.v. "World Oil Market Chronology from 2003," last modified August 25, 2017, https://en.wikipedia.org/wiki/World_oil_market_chronology_from_2003 (accessed August 29, 2017).

2. James Worrell, David E. Marshall, and Jon Fisher, "The Nitrogen Bomb," *Discover Magazine*, April 1, 2001, http://discovermagazine.com/2001/apr/featbomb (accessed August 29, 2017).

3. Ibid.

4. "All Nobel Prizes," Nobel Media AB, 2014, https://www.nobelprize.org/nobel_prizes/lists/all/ (accessed March 1, 2017).

5. Ibid.

6. "*Graphene: The Worldwide Patent Landscape in 2015*" (Newport, UK: Intellectual Property Office, March 25, 2015), https://www.gov.uk/government/publications/graphene-the-worldwide-patent-landscape-in-2015 (accessed July 20, 2017).

7. *Wikipedia*, s.v. "List of Edison Patents," last modified June 8, 2017, https://en.wikipedia.org/wiki/List_of_Edison_patents (accessed August 19, 2017).

8. Jill Jonnes, *Empires of Light: Edison, Tesla, Westinghouse, and the Race to Electrify the World* (New York: Random House, 2004).

9. *Wikipedia*, s.v. "Bayh–Dole Act," last modified August 24, 2017,

https://en.wikipedia.org/wiki/Bayh%E2%80%93Dole_Act (accessed August 29, 2017).

10. Richard Pérez-Peña, "Patenting Their Discoveries Does Not Payoff for Most Universities, a Study Says," *New York Times*, November 20, 2013, http://www.nytimes.com/2013/11/21/education/patenting-their-discoveries-does-not-pay-off-for-most-universities-a-study-says.html?mcubz=0 (accessed June 17, 2017).

CHAPTER 9: GRAPHENE IN SPACE!

1. D. A. García-Hernandez et al., "The Formation of Fullerenes: Clues from New C_{60}, C_{70}, and (Possible) Planar C_{24} Detections in Magellanic Cloud Planetary Nebulae," *Astrophysical Journal Letters* 737, no. 2 (August 20, 2011): L30.

2. In the case of the Voyager and New Horizons spacecraft, this time will be measured in millennia.

3. L. Johnson, et al, "Near Earth Asteroid (NEA) Scout," Fourth International Symposium on Solar Sailing (ISSS 2017), Kyoto, Japan, January 17–20, 2017, https://ntrs.nasa.gov/search.jsp?R=20170001499 (accessed November 14, 2017).

4. "Voyager," Jet Propulsion Laboratory, California Institute of Technology, last modified July 18, 2008, https://voyager.jpl.nasa.gov (accessed June 4, 2017).

5. Jacob Aron, "Spacecraft May Fly on Graphene Wings," *New Scientist* 226, no. 3023 (June 2015).

6. This laser ablation is a popular notion among Earth defense enthusiasts as well. Firing the same terawatt lasers, which would hypothetically power laser sails, could also be aimed at threatening celestial objects. The light would vaporize rock and ice at the point of impact, effectively turning the asteroid into a mini-rocket.

7. Genesis 11:4 (King James Version).

CHAPTER 10: GRAPHENE CYBERNETIC ORGANISMS

1. Emiliano Lepore et al., "Silk Reinforced with Graphene or Carbon Nanotubes Spun by Spiders," *arXiv*:1504.06751 [cond-mat .mtrl-sci] (2015).

2. Nihar Mohanty and Vikas Berry, "Graphene-Based Single-Bacterium Resolution Biodevice and DNA Transistor: Interfacing Graphene Derivatives with Nanoscale and Microscale Biocomponents," *Nano Letters* 8, no. 12 (2008): 4469–76.

3. Alessandro Fabbro et al., "Graphene-Based Interfaces Do Not Alter Target Nerve Cells," *ACS Nano* 10, no. 1 (2016): 615–23.

4. Hannah Devlin, "What Is Functional Magnetic Resonance Imaging (fMRI)?" Psych Central, June 1, 2017, https://psychcentral.com/lib/what-is-functional-magnetic-resonance-imaging-fmri/ (accessed August 29, 2017).

5. T. S. Sreeprasad et al., "Graphene Quantum Dots Interfaced with Single Bacterial Spore for Bio-Electromechanical Devices: A Graphene Cytobot," *Scientific Reports* 5, article number: 9138 (2015).

6. Yinfeng Li et al., "Graphene Microsheets Enter Cells through Spontaneous Membrane Penetration at Edge Asperities and Corner Sites," *Proceedings of the National Academy of Sciences United States of America* 110, no. 30 (2013): 12295–300.

7. J. D. Lanphere et al., "Stability and Transport of Graphene Oxide Nanoparticles in Groundwater and Surface Water," *Environmental Engineering Science* 31, no. 7 (July 2014): 350–59.

8. "A Realistic Assessment of Graphene Toxicity: An Interview with Andrew Maynard," The Graphene Council, July 2014, http://www.thegraphenecouncil.org/?page=GrapheneToxicity (accessed August 30, 2017).

CHAPTER 11: USING THE REST OF THE TABLE

1. "The Nobel Prize in Chemistry 2016," Nobel Media AB, 2016, https://www.nobelprize.org/nobel_prizes/chemistry/laureates/2016/ (accessed May 2017).

2. José Berna et al., "Macroscopic Transport by Synthetic Molecular Machines," *Nature Materials* 4, no. 9 (September 2005): 704–710.

3. Rotaxane rings are attracted to special places on the handle because of intermolecular forces designed into the three-part (handle, weights, ring) system. Only places on the handle that have been designed to act favorably with the ring are considered stops where the ring can exist for any measurable amount of time. Chemists can design as many stops as they would like along the handle's length, although two locations are most common when discussing binary "off/on" behavior within a rotaxane. All other locations along the handle react unfavorably with the ring, and the ring simply shuttles between stops when it has enough energy to make the jump.

4. Nagatoshi Koumura et al., "Light-Driven Monodirectional Molecular Rotor," *Nature* 401, no. 6749 (September 9, 1999): 152–55.

5. Matt Davenport, "World's First Nanocar Race Crowns Champion," *Chemical & Engineering News*, May 2, 2017, http://cen.acs.org/articles/95/i19/Worlds-first-nanocar-race-crowns-champion.html (accessed May 29, 2017).

6. Up, down, left, right, front, and back.

7. Remember, groups in the periodic table are the vertical columns. Elements with similar electronic properties are lined up on top of one another. Having similar electronic properties, they also react similarly. This is one of the key predictions underlying the periodic nature of the table.

8. Buckminster Fuller, "Nine Chains to the Moon," Southern Illinois University Press, 1963, Carbondale, Illinois.

9. That we have found so far. A geologist who discovered rocks with high-purity germanene or stannene within them would find themselves with many science awards, as well as financial security for life.

10. Yang Guo et al., "Superconductivity Modulated by Quantum Size Effects," *Science* 306, no. 5703 (2004): 1915–17.

11. Baojie Feng et al., "Evidence of Silicene in Honeycomb Structures of Silicon on Ag(111)," *Nano Letters* 12, no. 7 (2012): 3507–511.

12. Yongmao Cai et al., "Stability and Electronic Properties of Two-Dimensional Silicene and Germanene on Graphene," *Physical Review B* 88, no. 24 (2013).

13. Feng-feng Zhu et al., "Epitaxial Growth of Two-Dimensional Stanene," *Nature Materials* 14 (2015): 1020–25.

14. Pronounced "em-ex-enes."

15. Just as electrons buzz around on a graphene flake but have a harder time jumping between stacked flakes.

INDEX